THE NORTHERN ROCKIES:
A FIRE SURVEY

To the Last Smoke

SERIES BY STEPHEN J. PYNE

STEPHEN J. PYNE

THE NORTHERN ROCKIES

A Fire Survey

THE UNIVERSITY OF
ARIZONA PRESS
TUCSON

The University of Arizona Press
www.uapress.arizona.edu

Printed in the United States of America
21 20 19 18 17 16   6 5 4 3 2 1

ISBN-13: 978-0-8165-3351-0 (paper)

Cover design by Leigh McDonald
Cover photo: Smoke plume rising from the Selway-Bitterroot Wilderness, 2013. Courtesy of the
U.S. Forest Service.

Unless otherwise noted, all photos are by the author.

Library of Congress Cataloging-in-Publication Data
Names: Pyne, Stephen J., 1949– author. | Pyne, Stephen J., 1949– To the last smoke ; v. 3.
Title: The Northern Rockies : a fire survey / Stephen J. Pyne.
Description: Tucson : University of Arizona Press, 2016. | Series: To the last smoke / series by
    Stephen J. Pyne ; volume 3 | Includes bibliographical references and index.
Identifiers: LCCN 2016004470 | ISBN 9780816533510 (pbk. : alk. paper)
Subjects: LCSH: Wildfires—Rocky Mountains—History. | Wildfires—Montana—History.
    | Wildfires—Idaho—History. | Wildfires—Rocky Mountains—Prevention and control—
    History. | Wildfires—Montana—Prevention and control—History. | Wildfires—Idaho—
    Prevention and control—History. | Forest fires—Rocky Mountains—History.
    | Forest fires—Montana—History. | Forest fires—Idaho—History. | Forest fires—Rocky
    Mountains—Prevention and control—History. | Forest fires—Montana—Prevention and
    control—History. | Forest fires—Idaho—Prevention and control—History.
Classification: LCC SD421.32.R62 P96 2016 | DDC 363.37/9—dc23 LC record available at
    http://lccn.loc.gov/2016004470

♾ This paper meets the requirements of ANSI/NISO Z39.48-1992 (Permanence of Paper).

To Sonja,
old flame, eternal flame

CONTENTS

SERIES PREFACE

To the Last Smoke

WHEN I DETERMINED to write the fire history of America in recent times, I conceived the project in two voices. One was the narrative voice of a play-by-play announcer. *Between Two Fires: A Fire History of Contemporary America* would relate what happened, when, where, and to and by whom. Because of its scope it pivoted around ideas and institutions, and its major characters were fires and fire seasons. It viewed the American fire scene from the perspective of a surveillance satellite.

The other voice was that of a color commentator. I called it *To the Last Smoke*, and it would poke around in the pixels and polygons of particular practices, places, and persons. My original belief was that it would assume the form of an anthology of essays and would match the narrative play-by-play in bulk. But that didn't happen. Instead the essays proliferated and began to self-organize by regions.

I began with the major hearths of American fire, where a fire culture gave a distinctive hue to fire practices. That pointed to Florida, California, and the Northern Rockies, and to that oft-overlooked hearth around the Flint Hills of the Great Plains. I added the Southwest because that was the region I knew best. But there were stray essays that needed to be corralled into a volume, and there were all those relevant regions that needed at least token treatment. Some, like the Lake States and Northeast, no longer commanded the national scene as they once had, but their stories were interesting and needed recording, or like the Pacific Northwest or central oak woodlands spoke to the evolution of fire's American century in a new way. I would include as many as possible into a grand suite of short books.

My original title now referred to that suite, not to a single volume, but I kept it because it seemed appropriate and because it resonated with my own relationship to fire. I began my career as a smokechaser on the North Rim of Grand Canyon in 1967. That was the last year the National Park Service hewed to the 10 a.m. policy and we rookies were enjoined to stay with every fire until "the last smoke" was out. By the time the series appears, 50 years will have passed since that inaugural summer. I no longer fight fire; I long ago traded in my pulaski for a pencil. But I have continued to engage it with mind and heart, and this unique survey of regional pyrogeography is my way of staying with it to the end.

Funding for the project came from the U.S. Forest Service, the Department of the Interior, and the Joint Fire Science Program. I'm grateful to them all for their support. And of course the University of Arizona Press deserves praise as well as thanks for seeing the resulting texts into print.

PREFACE TO VOLUME 3

N JULY 2012 I conducted a two-week road tour of the Northern Rockies. A number of places of interest I had visited before and chose not to revisit. For the rest, I mostly knew what I wanted to see and just needed to place my feet on the sites and settle my mind on their meanings. One essay was written previous to the trip, and a couple after the draft was completed, as I felt the need to revisit and develop certain topics. Those who made my travels productive (and in some cases, possible) are acknowledged in the individual essays.

The region is an inspiring setting for anyone interested in wildland fire. It brought home how the big differs from the novel and the important. This is a place where bulk rather than diversity was defining, where fewer themes prompted similar responses. It seemed a place where one grand attribute was repeated, though with new meanings as cultural circumstances change. The upshot is that a larger region resulted in a shorter book. The primary surprise was how a generational theme emerged—that, I hadn't anticipated. I became doubly grateful that I had the opportunity to join the cavalcade of those who have encountered the Northern Rockies' big sky, big mountains, and big burns.

THE NORTHERN ROCKIES:
A FIRE SURVEY

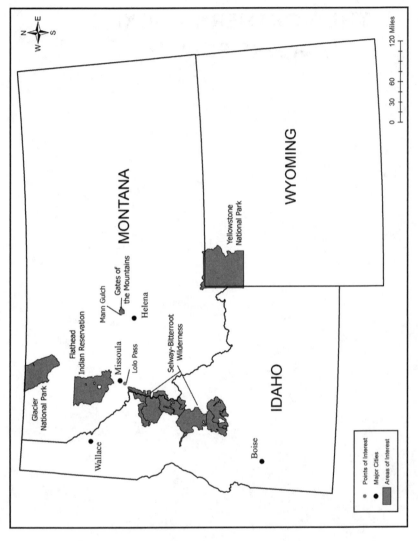

Map of Northern Rockies region. Map by Chris Miller.

PROLOGUE

Where the Mountains Roar

I T'S A BIG COUNTRY with big burns that have made big impacts on
the American fire community. At the onset of the 20th century the
national fire problem centered mostly on the Lake States, from New
York to Minnesota, where landclearing and settler burning had for 40
years routinely erupted into deadly conflagrations. But in 1910 the cultural equivalent of a passing cold front shifted the flames west. The Big
Blowup relocated the nation's fire problem from the raw scalped land of
settlement to wildlands, much of them not seriously touched by pick, axe,
or steam, and set aside as reserves before permanent marring occurred.
From then until the postwar era the big burns that commanded national
attention, grabbed agencies by their bureaucratic lapels, and shaped policy came with particular force from the Northern Rockies.

Other regions have had big fires, and in the decades following World
War II California has burned in counterpoint. But California's critical
conflagrations migrated to the urbanizing South Coast. The Northern
Rockies' blowups remained in the wild. The region's impact on national
policy and what might be termed the national style of fire management
has remained remarkably persistent. Federal firefighting began system-
atically in Yellowstone National Park when the U.S. Cavalry took over
administration in 1886. The Great Fires of 1910 traumatized the young
U.S. Forest Service (USFS), sparked passage of the Weeks Act that
laid down the infrastructure of a national fire protection system, and

encouraged the USFS to battle light-burning as a policy. The next three chiefs, through 1939, were personally on the firelines of the Big Blowup and used their fight against fires as a marker of success. The 1926 fires in Glacier National Park caused a fiscal crisis for the National Park Service (NPS), strengthening its determination to halt fires of all kinds. The 1934 outbreak in the Selways forced the USFS to grapple with the conundrum of big fires in its backcountry, and led to the adoption of the 10 a.m. policy in 1935. The smokejumper corps emerged in an effort to project power into the remote hinterlands. But big burns, and big fire busts that could overwhelm a force of smokechasers, continued. In 1937, with the Blackwater fire, the Northern Rockies announced a three-decade era of tragedy fires. In 1949 thirteen smokejumpers fell at Mann Gulch.

Big busts returned in 1961 in eerie alignment with the onset of the fire revolution. The Wilderness Act of 1964 jolted the region with special force since it held so much roadless area. The 1967 fires, particularly in Glacier, suggested both the mechanical limits of fighting such outbursts and the costs of attempting to do so. The next year the NPS had adopted a new policy to replace the 10 a.m. The Forest Service moved more cautiously. But in 1972 the White Caps of the Selway-Bitterroot Wilderness witnessed the first natural fire, the misnomered Bad Luck; the next year saw the first big natural fire, the Fitz Creek; and in 1988 a swarm of such fires sparked a second Big Blowup as they overwhelmed Yellowstone National Park and the Bob Marshall Wilderness, and unblinking media coverage brought the issue to the nation's consciousness. Free-ranging natural fires would be neither small nor inconspicuous: in politics, as with biotas, it is the big fires that do the work, and big fires the Northern Rockies have in abundance. The 2000 fire bust segued into the adoption of the National Fire Plan.

Its big burns are not the only fires of interest. Its montane forests, mostly of ponderosa pine, became textbook examples of the ecological rot that could result from ending fires in a landscape adapted to frequent surface burning. The Blue Mountains added a splashy demonstration of the forest health crisis, of which fire exclusion was a vital factor. The Yellowstone outburst showed the necessity of restoring high-intensity flame, not just understory burning. Its sprawl of exurban settlement, notably in the Bitterroot Valley, added to the national disaster slowly unfolding as an irrepressible burning met an implacable sprawl. While other regions became the poster child for each of these topics, the Northern Rockies gave them critical heft because of its importance to the Forest Service and because it had a fire culture that took fire seriously. The Missoula fire lab could broadcast nationally what happened in its big backyard.

For a long time Missoula was fire's company town, with a Forest Service presence and a cluster of fire institutions that included a lab, equipment center, smokejumper depot, and regional offices that made fire an assumed fact of life and discourse. The big burns then segued seamlessly into the wilderness movement, taking their place alongside ice-fringed mountain ramparts and grizzly bears as emblems of the Wild. They became a topic of academic discussion and seasonal gossip. They entered into literature; the modern genres of fire—fire as disaster, the firefight as a battleground, fire as tragedy—emerged out of the smoke of the region's big burns. Fire lore, or in regional terms, meeting fire in the woods, passed from generation to generation. What the Southeast became for prescribed fire, and Southern California for the I-zone, the Northern Rockies became for big burns in the backcountry.

The future suggests that the big burns are likely to become still bigger. Climate change will destabilize the northern region more than others. The big fires along the Bitterroots and the Big Belt will move in awkward fugue with the Big Burn of industrial combustion. The dialectic will be vigorous because it will lack the buffers of working landscapes, cities, and heavily humanized sites. Nature's pure fire will meet humanity's singular one. The Great Fires are still to come.

PORTAL

LOLO PASS

THE LOLO PASS breaches the Rockies southwest of Missoula. Here the fierce ramparts of the Bitterroot Range soften, the Lolo Creek creates an embayment into the Bitterroot Valley, and a path, often wide and slow, grades up from 3,200 feet to 5,225 at Lolo Pass. Over the pass the mountains, like an ocean of white caps, spill to the horizon.

It's an ancient portal, the traditional route by which peoples from the end of the Pleistocene to the age of steam passed through the mountains. It's where their guides led the Lewis and Clark expedition and so became the site where the irresistible force of American westering met the implacable barrier of the Rockies. That encounter moved the Lolo from a simple place into an epic narrative. But to pass over the ramparts of the Bitterroots is to enter a strange world.

The land was then, and has remained for another two centuries, a forbidding place both of bountiful scenery and of few means by which to live to enjoy it. The Northern Rockies were a place to pass through, not to linger within. The indigenes found trails to transit across quickly. The Nez Perce went east to the buffalo, the Salish west to the salmon. Lewis and Clark failed to find the river passage they sought. On their way west they suffered through cold and snow and scarcity; on their return east, they nearly starved before finally eating a colt. Stray trappers eked out a living on furs, but left no record other than blazed traplines and the errant log cabin. Prospectors failed to find gold and passed over the land like a placerless stream. Railroad tycoons failed to hack out a line over the granite crest and through the rugged forests, although the experience left them with checkerboard chunks of mountain for later exploitation. The land had trails but no farms, mines, ranches, logging camps, or towns. When the land was designated the Bitterroot Forest Reserve, its rangers, scarcely more abundant than trappers, found few natural resources to oversee. The region's natural treasures existed more in principle than in practice. What the valley of the Lochsa had was rock, water, sky, woods, and a silence that was the cultural equivalent of loneliness. Yet it was far from inert. It held blizzards, floods, swarming insects, wandering grizzlies, elk, and marten, and most dramatically, it had fires.

For 60 years after the Bitterroot Forest was proclaimed, the primary occupation of its inhabitants—rangers and smokechasers of the U.S. Forest Service—was to fight fire. That task demanded lookouts, trails, telephone wires, and guard stations. Those new occupants, seasonal as all humans in the Lochsa had always been, were the successors to a cavalcade of mountain men who had come to and engaged the land on whatever terms they could find. Their story, what defined them and their relationship to the land, retold like a Celtic saga, was the encounter with the wild. The landscape's burns became the latest point of contact, supplying new terms like *smokechaser, fire trench,* and *backfire* by which to fit the cadence of the old lays. Those fires financed a kind of ephemeral settlement, as furs had earlier. They were a means to manhood, the test by which a new generation would come of age. They compelled a physical engagement, not simply an aesthetic appreciation. The land had its ethic. Its fires tested newcomers against it.

When, with Civilian Conservation Corps labor, a paved road crested over the Lolo Pass, it paradoxically inspired and depressed the old-timers. Like many others they wanted easier access, since access would make available for exploitation what remoteness denied. But beyond livelihoods, they wanted a particular kind of life, and by changing the conditions of their engagement with the mountains, a road threatened to deny that way of living. Then, after World War II, as roads thickened into clear-cuts and spray fields, something snapped. The deep narrative of the Northern Rockies had not been about bold men opening up new lands for commerce: it had been about men testing themselves against and savoring the sustenance provided by the mountains. Those tests had involved axe, gun, trap, snowshoe, and the know-how to use them, not pushing a bulldozer or twirling the steering wheel of a logging truck. The furs and the fires had been a means to allow men to engage the mountains. Clear-cuts, terracing, and dozer-scoured logging roads were not the kind of relationship the thoughtful wanted. They sensed, in the words of one ranger, that they had "gone to the dark side."

So long as the only calculus of reckoning was economic, the old ways could not stand against the new. Then the wilderness movement gave alternative notions intellectual heft, political power, and legal standing, and the lands quickly segued into a formal wild for which the woods, the furs, and eventually the fires found a different purpose. Yet even as the lines of the story were being rewritten, the ur-theme, and its structure, remained as unyielding as the Bitterroot's batholiths. The hunter who put down his gun for a camera, the trapper who became an outfitter, set the pattern for the smokechaser who watched the fires burn.

A remarkable number of Forest Service chiefs, rangers, and fire officers have passed through the Lolo portal. They came to appreciate that the land—the wild—transcended any act they could marshal against it, not in a physical sense, since it was possible with modern machinery to scalp, scrape away, blow up, or otherwise permanently mar even the granitic substrate of the Rockies, but in a hard-to-verbalize moral sense. They realized that they were not the greater for overcoming the challenge of the wild but the lesser because they failed an ethical test. Eventually they appreciated that what they wanted was not the stuff they could haul out of the wilderness. It was the wilderness itself. Or more precisely, they wanted to be able to retell, or better, to relive, the story of what it meant to crest over the pass and gaze, wide-eyed and transfixed, on a prelapsarian world, and to pass that experience along to another generation.

THE MISSOULA MATRIX

"Mazoola," he was still grumbling. *"Why don't they send us to hell to study fire and be done with it? What I hear, the mileage is probably about the same."*

—RANGER VARICK MCCASKILL, IN *ENGLISH CREEK*

M ISSOULA IS ONE OF AMERICA'S great fire towns. Over the past few decades, particularly after the timber industry crashed and a service economy tried to replace it, the city has gentrified and suburbanized. Houses lap up the slopes of its converging valleys, boutiques have replaced brothels, university classrooms have risen while tepee burners for mill sawdust have been torn down, and a firefighting force has shifted from hard laborers pulled from railway or mining gangs to college-schooled technicians. But its taproots in the Forest Service and the surrounding wilds run deep, and for over a century that story has been written in big burns.

For the Northern Rockies the era of exploitation passed quickly, save for a few places like Butte where mining left suppurating scars, and swathes of the Bitterroots (and other production forests) where logging slashed through woods and bulldozed hillsides into terraces. Mostly the imagined treasures washed away rapidly, like placers; traffic blew through rather than planted; and the marring left by axe and pick melted away like the deep drifts left by a blizzard. What remained was a hard matrix of mountains and, beginning in 1891, an inland empire of forest reserves. The landscape was mostly roadless, and by 1964 much of it was destined for formal gazetting as wilderness or other forms of nature preserved. Unlike the Colorado Rockies, the wild mattered, not just as outdoor recreation. Unlike the Utah Rockies, or the Wasatch Range, which more than rivals the Bitterroots in scale, the Wild as wild resonated with public sentiment. The Northern Rockies kept their grizzlies—and they kept their fires.

Fire protection dominated early Forest Service administration: the Great Fires branded several generations. More than anything else, the perceived need for fire control "settled" the region. It put in the trails, scratched primitive access roads, strung telephone wire, erected lookouts and guard stations. Firefighting was a primary way of engaging with the backcountry. Missoula began amassing an institutional apparatus to match its burns. It had a major fire cache. The Remount Depot arose at Ninemile Ranger Station. In 1953, building on regional experiments in aircraft, the Forest Service created the Aerial Equipment Development Center at Fort Missoula, later expanded in 1959 to one of two Forest Service–wide development sites. In 1954 the service dedicated an Aerial Fire Depot and Smokejumper Center. In 1960 it opened a fire lab, one of three nationally, which subsequently interacted with the University of Montana (UM) Forestry School. When wilderness became an informing theme, the interagency Aldo Leopold Wilderness Research Institute was founded in 1993 in facilities of the Rocky Mountain Research Station located on the UM campus. An exceptional number of chiefs and fire directors have served in Missoula—and then retired there. When enthusiasts sought to charter a national museum to commemorate the history of the Forest Service, there was really only one site to consider: Missoula.

Yet buildings and bureaucracies do not a fire culture make. The hardware can only do what its software tells it, and what makes Missoula a hearth is not the stonework but the tended flames within.

Living fire cultures require a distinctive fire (displayed in abundance), an institutional infrastructure to engage it, and a supporting society to endow it with meaning. Firescapes are, in truth, cultural creations. It's not enough to have a physical setting with lots of fires or big fires. Nevada has both, and few people care because there is no narrative to engage it, no means for residents or visitors to see in those flames something that speaks to their identity, that moves fire from chemical reaction to evocation. The Northern Rockies matter because there is a poetry to match the mountains. In the same way that old-guard Floridians see prescribed fire not only as a tool but as a cultural legacy, so too can Montanans see their big burns as a valued part of their heritage. For the early fire community they were a defining feature of a vast backcountry; more recently, they are manifestations of a valued wild.

Missoula has its writers, and the mountains have a literature. The archetypal narrative, since Lewis and Clark invented it, is the encounter with a gigantic nature, either big in its mass or in its transcendence. A Montana school has repeated that story with new characters and new means of encountering, swapping beaver traps for, say, fishing rods, or the pack train for a Ford Trimotor, but the encounter as a means of coming of age endures from A. B. Guthrie and Norman Maclean to D'Arcy McNickle, Ivan Doig, and Willard Wyman. This is not a story inherent in the rocks, to be plucked out like veins of ore. It's made by mind working on mountain, as a comparison with encounters in other western landscapes shows. The Colorado Rockies invite amenities and outdoor recreation; it's an engagement by piton, bike, backpack, and ski. The California Sierras make encounter into a solitary quest; a finding of the self by a solo hike, a climb up a rock face, or the search for a personal epiphany. The Montana encounter, however, is generational. It involves a physical test, often unsought, imposed by a big countryside, but one that occurs under the tutelage of a senior, an old trapper, packer, or ranger, who seeks to pass along an understanding of the wild to another generation. It's an earthier story, one enlarging without becoming misty with diffused lights and one that moves beyond awareness of any one person.

It's a literature that requires the big wild as a setting; and for that reason it must accept fires as it does bears. The modern literature of wildland fire, to the extent that it exists either as event or as theme, has centered in the literature of the Northern Rockies. There is a popular acceptance, or at least acquiescence, that fire belongs, however begrudged when the long-smoldering backcountry can smoke in the Bitterroot or Flathead Valleys for weeks. It was possible here to transition from fighting big burns in the mountains to tolerating them, and even celebrating them, that is open to few locales, and none with so vast an estate in which to let them roam. That could not have happened without a public culture to support it, and without poets to inspire that public.

Today, the wild dominates the political geography of the Northern Rockies. Three massive clusters of legally wild land frame the region: one along the Selway-Bitterroot south into the Sawtooths; one in the Lewis and Swan Ranges; and one in the Greater Yellowstone Area. Add in, for USFS Region One, six National Wild and Scenic Rivers, the Continental Divide National Scenic Trail, the Lewis and Clark National Historic Trail, and abundant roadless areas and state parks and wildlife preserves. The classic assumption about wilderness is that it is an island amid development. In the Northern Rockies the wild is a lattice around which the rest entwines.

The number of "natural fires" and of acres burned by them is small compared to the full population, but still significant, and the experiment has demonstrated convincingly what can be learned about fire behavior and management by observing fire in something like its indigenous state (it's the difference between studying elephant behavior in zoos and in nature parks). The acres in wilderness will almost certainly rise. Even more, the experience with natural fire is allowing fire officers in areas outside wilderness to permit wildfires to burn more widely beyond the foundational wild. It's a second-order wilderness burning. To critics, it's a stealth imposition of the wild by means of fire, but to enthusiasts, it's a surrogate naturalness. Certainly it's less cynical than the logging-as-emulation-of-nature argument.

In the United States there are three fire culture regions, which together make a national fire triangle. Each region has a core, where understanding, institutions, and opportunities find a strong nuclear bond. That core absorbs the dominant character of its surroundings, transforms and invests sense in it, and then suffuses that hinterland with meaning that informs how to behave, which is to say, it proclaims an ethic about how to properly tend fire. If you want to learn how to do prescribed burning right, go to Tallahassee. If you want to see how wildland fires along an urban fringe work, go to Southern California. If you want to understand free-burning fire on a panoramic scale, or at least fire in wilderness, go to Missoula.

WHY BOISE IS NOT THE NATIONAL CENTER FOR FIRE

WHEN THE PUBLIC MEDIA want to know about fire, they might be excused for looking to the National Interagency Fire Center (NIFC) in Boise, Idaho. The name would seem to say it all. It's an official establishment. Every public entity associated with fire is present, from the Forest Service to the Bureau of Indian Affairs, from the U.S. Fire Administration to the Department of Defense; even the National Association of State Foresters has a representative. Many agencies house their national fire directors there. It gathers and posts the nation's fire statistics. NIFC is, so it might seem, where the action is.

In fact, it isn't. It's where the country orchestrates its logistical response to wildfire. It's not where the nation manages its fires: it's where it decides how to fight them. Having the national dispatch center in Boise is like having the world's highest tower in Dubai. It's a quirk of history. Worse, its visibility deflects attention towards emergencies and away from the slog of land management on which wildland fire's management will ultimately succeed or fail. Instead of fusing fire, as its name suggests, it shows fire's fissures. As the fire revolution boiled over, the architects behind NIFC opted to stay the course. It was a new means to old ends. It didn't bring the country to a new way of thinking about fire. It outfitted it to do the old job better.

Of course it does essential work. Fire management is impossible without the capability for fire suppression. If NIFC didn't exist, we would have to invent it. But its particular expressions; its quirks, beginning with its siting in Boise; its bid to be more than it is—all these demand historical explanations. Like a flamingo, which had to adapt a bill that appears upside down relative to its needs, NIFC at Boise is the product of historical accident. Like organic evolution, fire history must evolve out of what it inherits, not what it would like.

WHY BOISE?

When the Boise Interagency Fire Center was dedicated on July 25, 1970, Boise had little fire culture. It had no dedicated lab, no equipment development center, no logistical depot, no historic fires, no unique fire practices, no fire art, no fire literature, no special fire tradition. It had the Boise National Forest. It was nicely situated to respond to fires in the Great Basin, and while these were of serious interest to local ranchers and the Bureau of Land Management (BLM) who technically administered them, and while a cheatgrass infestation was primed to become a poster child for invasive pyrophytes, the fires' rumbling barely registered on national seismographs. Boise was an agreeable city, about to become a bigger one. Any objective observer would conclude that there was no reason to site a national center for fire suppression there.

What changed those calculations was the painful chrysalis of a modern Bureau of Land Management. When Jack Wilson was hired as a range conservationist by a BLM Wyoming district in 1948, the staff consisted of him (as manager) and a clerk to oversee 20,000 square miles. Fire was a seasonal, ancillary duty, fought by per diem guards, ranchers, nearby USFS personnel, and a 1943 bomb carrier converted to a fire engine. It was a situation "very typical at least of most BLM districts between 1936 and 1955." In fact, Wilson concluded, "I was little different than the GLO ranger in 1897."[1]

When fires broke out in 1961, the BLM had to fly in smokejumpers from Alaska—hardly a sustainable practice. When they roared around Elko, Nevada, in 1964, the agency was little better prepared than in Wilson's field career. A temporary dispatch center was established at Salt Lake

City. Still, under the dynamic leadership of Interior Secretary Stewart Udall, the agency was modernizing. It lacked a formal charter (it would not receive one until 1976), but its working code was to emulate the multiple-use practices of the U.S. Forest Service. To the extent possible it would be an Interior clone. If the Forest Service had a *USDA* prefix, the Bureau of Land Management would affix *DOI*. It would aspire to the same goals, seek to apply the same methods, and crave the same standing. That meant it had similar ambitions for fire protection, and some strategists like Wilson appreciated that fire was a fiscal mechanism by which to build up and out. After a postseason review, the agency established a permanent Great Basin Fire Control Center at Gowen Field. It recognized that Boise was well positioned to oversee its Intermountain Empire, and it knew that the Forest Service was independently interested in an interregional dispatch facility in Boise and was negotiating for a site. Cooperation led to a mutual Western Interagency Fire Coordination Center ready for the 1965 season.

If it appeared a daring ante for what seemed a weak hand, the agency did have a strong suit in its Alaska program. When the BLM was created in 1946, the executive order slapped together not only the old Grazing Service and Interior oversight for mining but also the Alaska Fire Control Service. As Alaska headed toward statehood in 1959, and as it assumed frontline standing in the Cold War, that tiny outfit grew quickly and went airborne. It learned how to run a modern, aerial firefighting program across a vast, sparsely settled interior basin between two bounding mountain ranges.

Quickly, the agency put its apparent greatest need—to prevent another outbreak like 1964—with its greatest asset, the Alaska fire operation. With a smidgen of imagination it was possible to equate the Yukon Valley with the Great Basin; to see in Boise a counterpart to Fairbanks; and to adapt a similar program of aerial fire control from taiga to sagebrush. Roger Robinson, head of Alaskan fire, transferred south to oversee the project. By now, however, the agency's ambition had gone national. The BLM wanted the Boise base to be its primary cache, dispatch office, and national ops center.

———————————

The BLM knew it could not do it alone. The Forest Service had promoted mutual aid since its origins, but mostly with state and private cooperators;

by the late 1960s it was slowly coming to the conclusion that it, too, could no longer cope by itself with the fire scene on the public domain. It was ready, region by region, agency by agency, to foster alliances.

In July 1967 the BLM and USFS signed a memorandum of understanding between themselves and the Weather Bureau for joint occupancy at a facility at Boise. Construction began soon afterwards, under BLM auspices. The facility went operational two years later. Each agency ran its own program: they just shared buildings and, where helpful, dispatchers. The warehouse had a yellow line painted down the middle to segregate each agency's cache. By now the entity was known as the Boise Interagency Fire Center (BIFC).

The negotiations clearly demonstrate that the two partners saw the project through very different prisms. For the Forest Service, BIFC was one among a suite of regional fire centers. For the BLM, which lacked the capacity for an equivalent array of facilities, BIFC was its primary fire center. More to the point, BIFC was its claim to national stature. It built the structures; it wanted its appointee to be identified as director. The Forest Service thought BIFC was a shared operation of equal partners. The BLM didn't want to share. It demanded that one person be in charge and that that person would wear a BLM patch on his shoulder.

The correspondence and minutes don't make uplifting reading. The Forest Service was reluctant to shed its dominant role nationally, though within four years it would publicly proclaim its commitment to a doctrine of total mobility. It saw BIFC as a "unique event," not likely to be replicated elsewhere and without special national merit. The Bureau of Land Management saw BIFC as its coming-out party. The BLM refused to accept any label for its appointee other than director. Back and forth the two agencies talked, fudged, finessed, and finally kicked the argument upstairs. By now the Forest Service was feeling the general upheaval of the fire revolution and finally conceded the matter of titles. The BLM representative would be called "BLM, Director," and the USFS representative, "Forest Service, Administrator." The memorandum of understanding was signed in May 1969. Chief Forester Edward Cliff spoke at the formal inauguration the next year. BIFC's first director was Roger Robinson; his successor, Jack Wilson.

As the Forest Service hegemony fractured further over the coming decades, like flakes spalling off a granite boulder, more authority for fire suppression seeped through the shards to BIFC. Additional agencies joined. In 1993 the facility changed its title to the National Interagency Fire Center. Yet it was ever more apparent that the National Interagency Fire Center did only one thing: it coordinated the countrywide response to wildfire. Its true core was the National Interagency Coordination Center, and its primary mission, to set priorities and arrange logistical support for going fires. It did not, as its old name and many partners might suggest, serve the nation's gamut of fire management needs.

⸻

When the fire revolution developed sufficiently that the nation's primary players needed to coordinate about other matters, a 1976 memorandum of understanding created the National Wildfire Coordinating Group. The group was strictly a talk shop to work through the particulars of what total mobility actually meant and how to implement it. It made sense to house the program at BIFC. But it could have been anywhere (and initially was, in DC). So, too, a dispatching center would have been anywhere (the Forest Service had set up an experimental operation in Alexandria, Virginia in the late 1960s). A cache needed to be close to the action; a dispatch center didn't. NASA had landed men on the Moon from Houston.

BIFC did not invent interagency cooperation: it just changed its terms and the relative powers of its members. It did nothing to change policy. Two years after its inauguration the Forest Service managed its first natural fire and three years later reincarnated its Division of Fire Control into a Division of Fire Management, advocated total mobility, and accepted the need for fire management to be grounded in land management. The BLM did nothing comparable. It pursued an old-model agenda for fire: the suppression system of the USFS during the 1950s and 1960s, the era when Interior's fire officers had looked covetously upon the Forest Service and seethed over their lower-caste status. NIFC made fire suppression more effective and more efficient and, in some respects, even glamorous. It was more a symptom of the fire revolution than an impetus for it.

The fire revolution that bubbled out of the Sixties had two prongs. One involved what was done, and the other, who did it. The problem with the American fire scene was not just that suppression was the informing policy, but that one agency, the U.S. Forest Service, had become a hege-mon. At an operational level they complement each other. While fire pol-icy was becoming more pluralistic, so were fire's institutions. The evolu-tion of BIFC can be seen as a forcing mechanism for that transformation.

The paradox, of course, is that even as the BLM pushed for institu-tional parity, made concrete in the facilities at BIFC, it was organizing its own fire program along the outmoded lines of the old Forest Service. It was promoting the kind of full suppression program that the fire revolu-tion of the Sixties had sought to redirect if not dismantle. BIFC became a vehicle to hold the old order. The BLM wanted to be an equal partner in that old order, not a vanguard of the new. In this respect BIFC was a high-tech throwback. It could have happened without the rest of the fire revolution. That it was planted in Boise only accented the extent to which it was removed from the hearths of wildland fire culture.

Boise is still an agreeable city with a large public sector economy, of which NIFC is an important part. Most fire directors much prefer to live there than around the DC beltway. But the city still has no fire research capabilities. Boise State University does not teach fire science—forestry and fire belong with the University of Idaho at Moscow. Idaho has five community colleges that offer certificates in fire science, none at Boise. It has no equipment center, only a massive depot. It boasts no organic fire culture, only a commitment to suppression that reflects a historic moment when an adolescent BLM decided to challenge an aging USFS amid a larger background of upheaval in how America saw fire on its wild and working landscapes.

Fire management requires the ability to fight fire. NIFC shows that ambition at a high level. The model, adapted, has gone to Winnipeg, Canada, and Guadalajara, Mexico. NIFC is where journalists and the public first look for information about the American fire scene. But if you want to know the fire story in the round, look elsewhere.

THE PARADOXES OF
WILDERNESS FIRE

PROLOGUE: FROM SUPPRESSED TO CELEBRATED

WHEN THE HAYDEN SURVEY probed into Yellowstone in 1871, they found their cross-country passage blocked by jack-strawed lodgepole pine left by fires. Extensive burns marred the scenery and made travel onerous. Yet the expedition's artist, Thomas Moran, painted its hot springs and woods and, spectacularly, its grand canyon; but not its burns. Partisans for reserving the place as a public park knew the Yellowstone's value lay in its display of Nature's wonders, and they believed these could be damaged by Nature's outbursts as well as by human vandals. The park's enabling legislation accordingly sought to preserve the place "from fire and axe." The origins of fire protection by the federal government, and its paramilitary style, commenced in 1886 when the U.S. Cavalry assumed control over the park. Fires greeted M Troop as they rode over the park boundary.

Over the next century the meaning of "wild" changed and "fire in the wild" with it. Initially, wild fire was, by definition, uncontrolled and unwanted. The more protected the scene, the more its overseers railed against free-burning fire. But by the time wilderness received its legislative mandate in 1964, a reconsideration was underway. Wilderness held a cluster of values, like an electron cloud. It was an outdoor gymnasium, a place of solace, a laboratory, and a baseline from which the relationship between humans and nature might be surveyed. At its nuclear core it was a place "untrammeled" by meddling humans, who might visit but would not reside. For most of the public it remained an untouched scene, still burnished from the Creation.

For an increasing fraction of the populace, however, wilderness did not seek to preserve that scene like an artifact in a museum, but aspired to retain its essential processes. In the philosophy of wilderness, process preservation replaced scenic preservation. If wolves, grizzlies, floods, landslides, and storms belonged, then so did fire. What unhinged wild places was not fire but fire's exclusion. Suppressing fire had no more justification than exterminating predators or channeling streams. The philosophical thesis for restoring fire was unanswerable, and unsurprisingly, the wilderness movement added powerful leverage to the fire revolution of the Sixties. On working landscapes fire might be useful, but in wild ones, it was mandatory. For the first 20 years of the revolution, beginning with policy reforms by the National Park Service in 1968 and ending with the 1988 Yellowstone fires, wilderness fire dominated national attention. The rest of the region's fire practices bathed in its reflected light.

More than a century after Thomas Moran painted his gorgeous canvases, there was scant argument among the informed over whether fire belonged. The issue was how to do it. In this discourse the Northern Rockies mattered because it held so much land, and because so much was managed with regard to fire as de jure or de facto wilderness. When the national fire community met in 1983 and again in 1994 to discuss wilderness fire, they convened in Missoula.

PARADOX 1: THE NATURAL AS CULTURAL

What has made a reconciliation tricky is that neither wilderness nor fire is a constant. They are variables, and while both have obvious natural properties, and while wild and fire can survive nicely without people, they are deeply embedded in human culture.

Fire has been a defining property for humanity since our origins as a species; nothing so clearly informs our ecological agency as our species monopoly over flame. Even to yield the exercise of our firepower willingly is a deliberate decision and one that typically has the threat of political coercion behind it. But wilderness, too, however ironically, is also

a cultural invention. It is not "natural" for people to remove themselves
from a landscape, and the American notion of wilderness, as distinct
from traditions of sacred groves or protected natural sites elsewhere on
Earth, is an expression of an exceptionalism that does not travel well outside the United States. Combine fire and wild, and the confusions multi-
ply. Each destabilizes the other.

A premise behind wilderness is that it stands outside culture. It can't.
It's more accurate to consider wilderness as more a state of mind than
a state of nature and to file the wilderness idea as an invention, one of
America's most singular. It had a long gestation. It was birthed during
the Romantic era and grew up in counterpoint to industrialism, the clos-
ing of the agrarian frontier, and a population migration from countryside
to city, all of which also coincided with spreading literacy and search for
nationalist symbols. Wilderness was what America had instead of cathe-
drals and coliseums. It helped that the country had vast swathes of nom-
inally empty land, a tabula rasa awaiting a text. Of course these lands had
not been vacant since the Creation: they had been continuously inhabited—cultured—for millennia; but the suddenness and violence of European contact had emptied them. To observers from afar they appeared
new, touched only by the grace of God.

There are plenty of examples of intellectuals inventing a lore for
a place (or a people). The Alhambra was just a ruin until Washington
Irving, then America's consul to Spain, wrote about it. The frontier trapper was a fringe figure, mostly comical, until James Fenimore Cooper
transformed him into the Leatherstocking Tales. A famous, if contro-
versial, example involves, in the words of Hugh Trevor-Roper, "the
whole concept of a distinct Highland culture and tradition" as "a retrospective invention." He argues that it was, in fact, a creation of the 18th
and 19th centuries, a time when the Highlands were becoming extinct
as an autonomous polity, when Highlanders were being removed from
the land, and when industrialization was replacing a folk economy with
a modern one. Instead, a combination of Romanticism, literary forgery,
commercial connivance, and the creation of a Highland Society among
elites conspired to "invent" a tradition. The Highland epic, the poems
of *Ossian*, was the 1760 fabrication of James Macpherson. The kilt was
"invented by an English Quaker industrialist" and "saved from extinction
by an English imperialist statesman."[1]

The history of American wilderness as an idea maps so neatly onto this scenario that it's surprising that the Big Trees weren't draped in tartan bunting and Mammoth Hot Springs serenaded with the "Braes of Lochiel." The authoritative survey, Roderick Nash's *Wilderness and the American Mind*, is a self-proclaimed history of an *idea*, one that came to a handful of poets and prophets and then propagated throughout the culture and became embedded into the national narrative, was coded into law with the Wilderness Act, and is valued for its celebration of an American identity.

PARADOX 2: THE WILD AS NARRATIVE

Americans have generally subscribed to a version of Scottish common-sense realism that holds the world outside our minds is real. With or without people nature goes on, and with or without a notion of wilderness, there exists a natural realm that has, can, and will continue to function on its own. In recent decades various nonanthropocentric philosophies have appeared to argue for the intrinsic value of that world and to plead for its preservation.

These notions can assume various national forms or intellectual traditions. Parks Canada sets its goal as ecological integrity, not naturalness or a historic condition. The Soviet Union established *zapovedniks* as pristine sites available for research. Most societies have some variety of sacred grove. Even in the United States a commitment to biodiversity may trump other values and require active intervention. American nature finds protection in many forms for many reasons, and these may not—often do not—agree in their particulars.

What holds together the American notion of wilderness—the strong nuclear force that keeps a buzzing cloud of values in orbit—is a narrative that makes wilderness part of a founding myth. The United States, so this saga goes, is the outcome of the encounter by European civilization with American nature. The early conversion of the wild into the cultivated traces the narrative of American progress. The more recent preservation of the wild as wild testifies to the maturity and triumph of that national narrative. Wilderness, in brief, is not something marginal or decorative to American identity. It is what makes the dialectic by which America has evolved. In place of monuments from antiquity America has marvels of nature. Wilderness matters ultimately because it explains who we are.

Wilderness is also a historical construct. When and how it appeared as landscape granted legal definition reflects certain times and conditions of the national experience. With eerie parallelism the modern wilderness movement is bracketed by the country's demographics. The first primitive area in the national forest system, the Gila, was proclaimed in 1924, the year the restrictive Immigration Act took effect, which yielded the lowest and most selective era of newcomers in American history. Both could be seen as acts of assimilation, of stabilizing a sense of American identity, while the country weathered the Great Depression, World War II, and the onset of the Cold War. The modern era of immigration, the largest by number in American history, commenced in 1965, the year after Congress passed the Wilderness Act. Simple coincidence? Possibly. But both emerged from a shared culture, and surely such nativist movements as See America First bespoke a shared sense of common context and challenge. Wilderness, in brief, was not just an idea bonded to American experience but to a particular era of American experience in which the nation looked inward even as it confronted global crises.

This helps explain why the concept has traveled poorly outside America's borders. National parks, biosphere reserves, wildlife refuges, national forests, even private lands held for public goods (such as the Nature Conservancy) have all circulated around the globe with relatively modest adaptations and concessions to local norms. Wilderness has not. It tends to be too absolute in its segregation of the natural and the cultural, and it is too closely bound to a particular narrative of American experience. If one wants to protect values in nature, or promote ecological goods and services, there are other concepts and practices possible; and wilderness may not be the best means to advance such goals. What grants it particular cultural traction in America is the narrative that has evolved to justify it—and this narrative is what halts it at the nation's frontiers.

There is nothing to prevent similar cultural alienation from happening within the United States as immigration again overflows, as the origin of immigrants shifts from Europe to Latin America and Asia, and as the newcomers do not head for geographic frontiers but to social or economic ones. The old story of wild America confronting civilized Europe no longer makes sense except as a historical moment. If what bonds people and land is culture, what shared culture allows for wilderness to continue to exert such force? The United States is becoming ever more multiethnic and pluralistic. The wilderness idea can no longer embrace its multitudes in a simple narrative.

Paradoxically, however, it is the putative acultural quality of the wild that may allow it to thrive. Critics have often noted that the presumed lands beyond the frontier were not genuine wilderness but homelands to native peoples. In the classic revisionist phrasing, these were not virgin lands but widowed lands. The wild was created by forcibly vacating the people who had resided there. Yet for a multiethnic nation, the standard formula for nationalist homelands—blood and soil—cannot succeed. The only land that all parts of a syncretic society can accept is one that is shorn of those ties.

Interestingly, wilderness has met with the greatest resistance where, in the national estate, the notions of blood and soil, of a people bound by a separate story of an ancestral or promised land for which they have suffered and endured, is strongest; in the South, in Texas, in Utah, on Indian reservations. Wilderness as a legal landscape works best where no population can claim it as a homeland. It works as a national commons precisely because it is not tied to any single group or ethnic-frontier narrative. It has the appearance of transcendence.

To endure, wilderness will need a philosophy, an ethics, and a story that can speak to all, and to make any practical difference in how Americans actually live on their land it will have to work on the ground. Wilderness might succeed as a new ethos, but it did not come with an operating manual.

PARADOX 3: THE WILD AS MANAGED LAND

Specifically, wilderness has to work within the agencies that manage fire on public lands (there is no legal wilderness outside the public domain). And it has to devise routine tools, a standard operating procedure, that will make overseeing fire in wilderness as much a part of the American fire scene as suppressing fire in the I-zone or burning off the rough of a Florida wildlife refuge.

America's public domain is a patchwork of purposes and agencies that has evolved over the course of the 20th century. The usual formula was, as Congress created new categories of land use, it identified an agency to administer them. The lands came first. The first national park, Yellowstone, had a hapless civilian superintendent; then the U.S. Cavalry took over the parks; and finally the National Park Service was created to oversee them in 1916, 42 years later. The forest reserves began in 1891, received an organic act in 1897, were run (more or less) by the General Land Office until in 1905 when responsibility was handed over to the Bureau of Forestry, which renamed itself the U.S. Forest Service. The first wildlife refuge was set aside in 1901, and given to the Biological Survey to administer; the Fish and Wildlife Service came into being in 1940 after a whirligig of reorganizations. The Taylor Grazing Act of 1934 closed the public domain to further privatization; the unpatented lands organized themselves into Grazing Districts; and the Bureau of Land Management was created by administrative fiat in 1946 and received a proper organic act in 1976.

But wilderness was different. The Wilderness Act established a distinct category of land with its own prescriptions for what can and can't be done, but it left designated wilderness sites within whatever agency held the original land. Each of the federal land agencies held wilderness; and since the Forest Service had so much land, and threatened to alter its roadless backcountry more gravely than the Park Service, so it received the most public attention, and drew the greatest controversy. The act did not create a National Wilderness Service to bring a common sensibility and protocol to the archipelago of the legally wild. (A compromise was to create an interagency Arthur Carhart National Wilderness Training Center and an Aldo Leopold Wilderness Research Institute.)

In one respect, letting agencies administer the land made a wilderness designation easier for a bureau to accept since it would not "lose" those lands. But it also meant that the agencies had to absorb internally the tensions between what its founding charter (and evolved identity) urged and what wilderness accepted or shunned. For agencies like the Forest Service, committed to multiple use and the norms of professional forestry, the strains could lead to ruptures within as well as rifts between the agency and the public. In the wilderness movement the Forest Service resisted, stalled, and generally put itself on the wrong side of history. To the new environmentalism that in the postwar era replaced old-style conservation, the Forest Service became the Evil Empire. The fight over wilderness alienated the agency; and the internal stresses threatened to tear it into pieces.

PARADOX 4: WILDERNESS FIRE AS MANAGED FIRE

Absorbing wilderness also caused internal schisms in field operations. Managing wilderness fire requires a different set of skills, temperament, and culture—a tradition that also had to be invented. Yet fire suppression still laid down the basic infrastructure. The same organization that had for decades committed to putting fires out as aggressively as possible would now have to let some of them burn. It was easier to accommodate prescribed fire since controlled burns put crews in the field, relied on similar gadgets and tactics for burning out, and often had *pre*suppression as a goal. A true wilderness fire program required extensive recoding in fire management's software.

Reform was not simply a question of principle but of practice. Although wildland fire had a place in the bureaucracy, it was not a bureaucratic category. It could not be programmed, budgeted, or predicted in the way timber harvesting or animal grazing units could be. It was, rather, an anticipated emergency and something to be fought, suppressed, and if possible excluded. At the time of the Wilderness Act fire control officers were using all available means to contain fire, including bulldozers, engines, air tankers, smokejumpers, hotshot ("shock troop") crews, and local agency militias. Along the fireline the Cold War met the backcountry. All this, however, violated both the spirit and language of the Wilderness Act. Natural fires belonged; as a matter of policy, it was wrong to expunge them. Ideally, fires should be left to free-roam. And when unwanted fires occurred, as they would when started by people or when wild fires threatened to cross borders, they had to be contained without recourse to the mechanization and heavy-boot-on-the-land that characterized the firefight as battlefield.

In some places, the Sierra parks most notably, parallel organizations were formed, one to fight fires and one to light or monitor them. Mostly, one organization had to learn to do both. It had to deal simultaneously with fire monitors as well as smokechasers, and branching decision charts in place of a simple directive for suppression. It had to learn when and how to tolerate fires that burned over seasons, not until 10 a.m. the next morning. It had to learn to tread as lightly as possible, or suppressing would scar the land worse than burning. This was a wrenching transition, like demanding a military engage in nation building instead of crushing an enemy. Wilderness fire caused as much internal tension as the wildland-urban interface, which forced wildland fire agencies to absorb the terms, tactics, and ethos of urban fire services. The Forest Service made its own transition worse by mulishly insisting it had to read the Wilderness Act with maximum literalness, like a teenager deciding to follow instructions in such a way that he hopes to show their fundamental foolishness.

Principles and practices were not easily segregated. Wilderness, after all, was not a scientific or logical idea but an evocation that spoke to a people's identity and hence to an ethic that embraced the land they shared. Yet apart from whether wilderness could be parsed from culture, there was the conundrum of people as the planet's keystone fire creature, as an integral chain in the natural history of being that had shaped wilderness for millennia. As in principle proponents had to deny wilderness as a homeland, so they had to deny anthropogenic fire as a process shaping those landscapes, even though removing the human presence might be as powerful as stripping away alpha predators. Even wild places—save the most remote and barren—had coevolved with humanity and its fires. Abolishing anthropogenic fire might be as disruptive as suppressing natural fires.

There was no neutral position possible. Overseers might choose not to hack out roads, log off woods, or dam streams. But they might need to burn. By the time wilderness fire came to dominate the fire community, everyone agreed that there was a lot less fire than in the past, that a lot of those lost fires had been good fires, and that this loss did ecological harm. Were those missing fires ignitions that nature had set and people unwisely suppressed? Or were they fires people had once set and no longer did? Such questions were neither trivial nor obvious. They required a sustained commitment that could only come from culturally compelling beliefs.

The outcome of not burning might be wild, but it might bear little relationship to the mix of natural features that had defined the land when Europeans first contacted it. On the basis of ecological goods and services, such as biodiversity, there was a good case to deliberately intervene and burn. But if people were justified in burning, then why not other practices as well?

Principles and practice could not readily be teased apart. Legal wilderness had at its nuclear core a paradox, that people must manage what is, by definition, autonomous from human meddling. For many practices it was possible to finesse around that paradox. It was not possible with fire.

PARADOX 5: THE PRAGMATIC EVALUATION OF THE WILDERNESS IDEAL

Nearly four decades of experience suggest both the power and the limits of wilderness fire. What seemed obvious in principle has blurred in practice as cultural ideals met natural realities. What seemed easy—simply standing aside and letting nature determine the outcome—has proved hard in practice. But the contours of fire and wilderness can now be rudely mapped.

It is clear that wilderness fire, which is to say, allowing natural ignitions as much autonomy as possible, works best in remote, self-contained landscapes—places with abundant natural ignition; landscapes with lots of room to maneuver, both for flame and smoke; and a local culture that favors the wild and is willing to make concessions to accommodate it. Where the wild is embedded within a larger domain of public wildlands, the local culture may be mostly the agency. And it is obvious that natural fire programs don't just happen: they work best when given hard leverage, when they become the default setting, through legislation such as the Wilderness Act or Endangered Species Act. If the program is to flourish, managers must justify a decision to suppress rather than a choice to allow to burn, which is less a choice than standard operating procedure.

Similarly, the limits of wilderness fire have become apparent. Size is a critical consideration. Wilderness areas can be as small as 5,000 acres, which doesn't allow much room to promote natural burns or to protect against outside fires, which in recent years are 20 to 100 times larger. The land will burn, and needs to burn; the issue is how, and with what consequences, or whether it is within the scope of the historic fire regime. Location (location, location) is equally critical. Escapes include smoke as well as flame; proximity to an urban area can be lethal to a program that tolerates large or long-smoldering burns.

The deeper challenges pertain to wilderness as a cultural construction. The legal wild is only one amid a pluralism of nature protections; the choice is not simply between the wild and the wrecked, but among a constellation of reserves, parks, refuges, and working landscapes dedicated to ecological goods and services. Each must seek a fire and management regime suitable to its purposes. As experience with wilderness matures, America will likely discover that the wild is not a place where ecological goods converge, that it involves a choice among the values a society wants from its lands. The likely outcome points toward a pluralism of fire practices, that wilderness will be one alternative among many, not simply the polar opposite of the urban. Probably the country needs less to recover its wilds than its middle ground, needs its working landscapes remade to promote values other than the maximum production of commodities and extractives.

To many partisans the idea that wilderness has a history will seem odd, if not offensive. The wild is transcendent—that's its essence—and hence lies outside history. The only history relevant to the wild is how people have over time come to understand it better as a fixed concept, not that the idea has itself evolved. But history runs all through wilderness, like quartz veins in granite. As an insight wilderness grew out of a particular time, place, and people; as a legal entity and a historical event, it helped shape an era of American fire. Very likely the fire revolution could not have happened without the fulcrum it furnished because it nationalized fire restoration as a project, unshackling it from a southern ghetto. The fact that it changes with its times makes it less a constant like the speed of light than a variable like interest rates.

Paradoxically, the wild is not something just *there*. It is put there by our engagement with it. A hard concept to grasp, perhaps, and a harder one to realize on the ground.

EPILOGUE: ALDO LEOPOLD AS EXEMPLAR

Over the course of a century and a half, wilderness has attracted many philosophers, publicists, prophets, and in recent decades, practitioners. But since the publication of *A Sand County Almanac* in 1949, it has increasingly bonded with Aldo Leopold. His fire biography is worth pondering.

He began his career as a conservationist at the Yale School of Forestry where he submitted an essay condemning bad logging and the slash fires it sparked. In 1913 as supervisor of the Carson National Forest he exhorted his staff that "we can at least assert that fire prevention is the most direct of all our activities, and hence also susceptible of developing the greatest relative efficiency." In 1920, after a decade with the

U.S. Forest Service in the Southwest, he staunchly upheld the agency's commitment to fire exclusion: "The Forest Service policy of absolutely preventing forest fires insofar as humanly possible is directly threatened by the light-burning propaganda." Over the next two decades he began to soften his approach to fire. Trashy logging, not fire, was the primary culprit in the North Woods; overgrazing, not burning, was the primum mobile in the Southwest. In the prairies of Wisconsin and the Chihuahuan Sierra he saw places that regularly burned without obvious damage. He corresponded with Herbert Stoddard, a pioneer in controlled burning and wildlife conservation in the Southeast.

But he never made an argument that fire belonged or that it might be necessary to burn in wilderness, or to argue against fire suppression as he did against predator eradication. He never wrote a "red fire" in the woods essay to match his "green fire" in the wolf's eye one. In 1925, a year after he successfully lobbied for the first Primitive Area in the national forest system, he lectured that "it would be idle to discuss wilderness areas if they are to be left subject to destruction by forest fires." The sand counties of *A Sand County Almanac*, where he attempted to rehabilitate abandoned land and had his famous "shack," was stripped to sand by ruthless cutting followed by deep burning. Fire had been part of the problem. It was not obvious how it might also be part of a solution. Fire was complicated.[2]

A history of conflagrations had branded itself into the consciousness of the founding generations of American conservationists. Leopold joined the Forest Service a year before the Big Blowup; fire control was a mandatory and accepted duty of his years as a ranger. He died when a fire sprinted across a neighbor's lot and, attempting to halt it, he suffered a heart attack, fell to the ground, and was burned over. His notebook bears the scorch marks.

So, after a fashion, do his ideas. Even in the wild, fire's management is a concept easier to work in the mind than in the hand.

FIRE'S CALL OF THE WILD

And from the fortieth winter by that thought
Test every work of intellect or faith,
And everything that your own hands have wrought.
— WILLIAM BUTLER YEATS, "VACILLATION"

O N JULY 2, 2012, a Cessna 182 climbed off the tarmac at Hamilton, Montana, for a fire recon flight over the Selway-Bitterroot Wilderness (SBW). That morning homeowners in Colorado Springs were being allowed back to neighborhoods incinerated by the Waldo Canyon fire, the most devastating in Colorado's history; that afternoon, an Air Force C-130 Hercules dropping retardant on a wildfire in the Black Hills crashed, killed four crewmen, and seriously injured two others; and large fires were ripping up parts of Montana to the south and east. The previous day lightning storms had passed over the Selway, and fire officers worried about fireworks on the upcoming Fourth of July, although errant bullets, careless campfires, and nature's own pyrotechnics were enough to keep the land aflame.[1]

But the Cessna observers were not looking for new smokes. They were tracking the enduring aftermath of old ones. They were on a fire recon through history. Forty years previously the first natural fire in U.S. Forest Service history had been allowed to burn above Bad Luck Creek. One of the two leaders of that brash experiment, Bob Mutch, was on board, as was Dave Campbell, ranger of the West Fork District of the Bitterroot, whose domain embraced that patch of the Selway-Bitterroot Wilderness. Their loop west and back was a journey across time written into geography. The land below bore the proud insignias of four decades of prudent aggression to reinstate natural fire in one of the storied landscapes of the American fire scene.

If Americans had a National Register of Historic Places for fire, the Selway-Bitterroot region would rank among the early entries. The mountains had been part of the Big Blowup of 1910. They had hosted the 1932 pack trip in which the leading lights of the Forest Service debated over evening campfires how to cope with big burns in the backcountry and, without accepting let-burning or loose-herding, had proposed minimum suppression. Then, the Pete King-Selway fires of 1934 had prompted a formal review of Forest Service policy that led to the promulgation of the 10 a.m. policy. Having battled helplessly against both historic conflagrations, Elers Koch of the Lolo National Forest argued that fire suppression across the Bitterroot summit had done more harm than good, and the agency would be better off letting fires burn and not tearing up the land with roads to improve access for firefights it couldn't win. But the struggle went on and a paved road punched over Lolo Pass and by the 1960s others began to appreciate Koch's prophecy. The Bitterroots became a national war zone for the protest against destructive logging those roads made possible. Despite improvements in firefighting and a smokejumper base at Missoula, fires, occasionally big fires, continued.

=====

What changed the calculus was the Wilderness Act of 1964. The Selway-Bitterroot Wilderness was among the first slated for inclusion. Still, someone had to connect wilderness with fire. The 1967 fire bust sent smokejumpers into the backcountry and bulldozers through Glacier National Park. Bud Moore, then on a fire inspection tour of Region One from the Washington office, reaffirmed the value of hard initial attack: "The strength, speed and effectiveness of Region 1 initial attack and first reinforcements actions needs study with the objective of identifying all possible means to improve them." Shortly afterwards, he returned to Missoula as regional director of Fire Control. Yet a sense gnawed at him that the old ways couldn't continue in fire any more than in logging, that "we had taken the shovel and pulaski folks about as far as we could." By 1970 he had translated a conviction that fire belonged in natural systems into a plan to test that vision. The restoration of fire might even seem an act of penance for what he had unthinkingly done to mar the Lochsa. Besides, granted the language of the Wilderness Act, he concluded that suppressing fires in such places was "almost illegal."[2]

He sought to identify possible sites, selected two, and then lit on the White Cap Creek area of the SBW, near where he himself had first experienced the wild. He found allies in Bill Worf, regional director of recreation, and Orville Daniels, supervisor of the Bitterroot, who agreed to host the trials. Then he recruited an agency planner, Dave Aldrich, and a fire researcher, Bob Mutch, to translate notions about the proper role of fire into formal documents that a bureaucracy could act on. In 1971 the White Cap Five hiked into the SBW and met over campfires, an eerie echo of the 1932 pack trip; they completed their task in the early summer of 1972. By then Sequoia-Kings Canyon and Yosemite National Parks had experimented with "let-burns" and Saguaro National Park had invented the concept of a "natural prescribed fire." On August 17, 1972, Chief Forester John McGuire signed off on the White Cap Project as an exception to the 10 a.m. policy. On August 18 lightning kindled a snag on south-facing slopes above Bad Luck Creek within the approved zone. Mutch and Aldrich scrambled to the scene but no one raised a shovel against the smoking snag, which smoldered in a brush field that claimed the site of an old burn. Fire on fire—the old fire changed the conditions for the new one. The Bad Luck fire burned an estimated 648 square feet.

The test came the next year when the Fitz Creek fire started on August 10. This one spread, creeping and sweeping in spasms, and it sparked a similar rhythm of responses. On August 13 and 14, as the fire moved uphill toward the Bad Luck Lookout, Moore ordered a B-17 to drop retardant to shield the facility. When lightning kindled 85 fires on the Nez Perce National Forest, the plume of the Fitz Creek burn attracted a crew, which assumed it was part of the bust, and Daniels had to intervene to pull them away. On August 18 the fire spotted across White Cap Creek, putting it outside the approved zone. In half an hour the escape, now named the Snake Creek fire, bolted up the slope, an independent crown fire, to the ridgetop. Daniels ordered an overhead team to suppress the escape (it was Mutch's team, for which he served as fire behavior analyst). But Daniels mandated control only for the Snake Creek fire, not the Fitz Creek. One fire could call out two responses. The pluralism that would characterize the fire revolution had its microcosms as well as its macrocosms.

Before it ended the Fitz Creek fire ran to 1,200 acres, found a chronicler in Don Moser for *Smithsonian* magazine, and could claim bragging rights for the best quote of the new era. It came from a red-eyed crew boss called in for the containment: "I'd like to meet the son of a bitch who said we had to let this one burn." Since the White Cap Project had an approved plan and had followed it, the Washington Office backed it up. The next year the Forest Service retitled its Office of Fire Control to Fire Management. Bud Moore retired.

The program had what it needed: a celebrity site, anointing by a Grand Old Man who had himself risen from the ranks, a stiff wind from the intellectual climate of the times, a trying demonstration, timeless anecdotes, room to expand, and the raw stuff for a legend. The Sierra sites were self-contained, and often overshadowed by the story of fire in the Big Trees. The Gila could blow up and no one might notice. Whatever the SBW's storied past for the fire community, the ongoing controversy over logging assured that the Bitterroot would be in the public eye. Its smoke plumes would be seen. It might even be that the naturally burned land might compensate for the mechanically scalped.

━━━━━━

That was how the program ignited. To propagate, it needed room to spread and authorization to do so. Both took time and a stubborn patience. Suppression remained the reflex response, the default option, the fallback or rally point when times turned tough. There was, in brief, nothing natural about natural fire. It happened because fire officers and district rangers chose it.

While the mountains might have a geologic unity, they were divided between two states, four national forests, and two Forest Service regions, all proud of their separateness and all granted considerable discretion to act differently. Expanding the land base, however, particularly through the 1981 Central Idaho Wilderness Act, put wilderness next to wilderness and nudged neighbors into common practices. After the 1988 Yellowstone brouhaha, every forest had to rewrite its fire plan, which encouraged a push toward common solutions along shared borders. The 1995 federal wildland fire management policy helped legitimize natural fire as an option. The program spread from the legally wild to adjacent wilderness study areas, roadless areas, and even general forest wildland, where fire behavior challenged survey lines.

The number of acres burned scaled up accordingly, never as much as had been taken out during the previous half century, and mostly keeping up with the expanding land base and branding the land in distinctive ways. There was enough fire to keep the program's momentum high and enough success to keep it pointed in the right direction. The numbers back up the anecdotes. In the decade preceding the Bad Luck fire, when wilderness ignitions were still being fought, the combined Selway-Bitterroot Wilderness and Frank Church-River of No Return Wilderness burned 4,989 acres. In the 1970s that figure scooted up to 6,838 acres, as a few fires began to take larger bites of the countryside. During the 1980s the figure increased fivefold to 38,026 acres. In the 1990s it doubled to 78,981 acres. In the first seven years of the new millennium it swelled half again as much to 122,982 acres. Suppression shrank to point protection and loose-herding along the redline borders that wildfire could not be allowed to cross.

What mattered were the large fires, or the big seasons, which held most of the big burns. They did the ecological work, and they began to rack up enough acres that successor fires burned into them in ways that tweaked the larger landscape into a kind of self-regulation. The administration adapted accordingly. Where, at Fitz Creek, a 1,200-acre burn had strained equipment, attention, and political will, by 1998 the SBW was accommodating multiple big burns; they reckoned they could handle 10 to 15 at one crack. The problem was negotiating outside the lines—with neighbors, even where the land was also wilderness, with nonwilderness national forest, with private lands overflowing with houses, with smoke in the deep valleys.

Then the contemporary cycle of big burns set in. While elsewhere in the Rockies suppression forces struggled to counter bust after bust, the West Fork District accepted more and more fires at once, subjecting them only to surveillance, and by so doing liberated scarce resources for more threatening wildfires elsewhere. The 2000 season shattered any lingering complacency about the power of suppression to contain wild-fire in the mountains; after 90 years the big burns still rambled more or less at will. But it also granted the new era an image of perhaps equal power to R. H. McKay's iconic photograph of the Nicholson adit where, at the height of the 1910 firestorm, Ed Pulaski had held his crew at gunpoint. John McColgan's *Elk Bath* photo from the Sula Complex in the East Fork of the Bitterroot National Forest suggested a fire as much at ease in the woods as elk. Whether formal wilderness or not, the East Fork was coming to resemble the West Fork. The argument that the fire management practices of the wilderness might be injected into general wildland gained a powerful visual logic.

A climax came in 2005, the centennial of the Forest Service. The West Fork District accepted 50 fires that season in the SBW. At one point four large burns were visible through a single pair of binoculars from the Hells Half Lookout. Managers relied on lookouts and webcams to monitor the scene. The fires began to merge, new burn into old burns, to break up, smolder, reform, blow up, creep into extinction. The complex went on for 60 days. When the season finally ended, Dave Campbell, who had been on the district since 1996, received a Chief's Award for Wilderness Fire Management. The next year he received another for Line Officer Wilderness Leadership.

Over the course of a century the Selway-Bitterroot Wilderness had gone from a place where wild fires threatened the political existence of the Forest Service to where fires, left to roam in the wild, promised its redemption.

Each season seemed to confirm the principles that, like some bureaucratic pyromancy, had been divined in the flames of Bad Luck and Fitz Creek. Or as that dour philosopher Arthur Schopenhauer once put it, "The first forty years of life give us the text: the next thirty supply the commentary." By the 40th anniversary the commentary of the fire program was being sculpted into the SBW's conifers.

One was that a fire program had to function as a system. It could not bounce from one emergency to another. It required a lot of fires over a large landscape bounded by similar practices on similar lands. A second was that one fire could be treated many ways. Some parts could be accepted, even applauded, even as other parts might need herding or outright suppression. A third was that, to restate Yogi Berra, sometimes you can see a lot just by looking. During the Fitz Creek blowup, Dave Aldrich blurted out that he had "learned more in a few days watching that fire than I did in the preceding 15 years fighting fire." If you want to manage free-burning fire, you have to understand how such fires burn, and you don't do that with your shoulder pinned to a shovel. You have to watch, ponder, and watch some more.

There is little doubt that the development of the Rothermel fire model, published the same year that the White Cap Project received official approval, suggested to wary administrators that they would now have the power to predict fire behavior with mathematical rigor and that those forecasts could undergird the decision to let natural fires roam. Edgy rangers could back up decisions with the seeming imprimatur of physics. Yet such forecasts were far beyond the model's capabilities (and still are). The only valid empirical basis had to come from the land itself, not a lab. It had to come by watching real fires, and that meant allowing real fires to run over woods and shrubfields, creek bottom and ridgeline. Even with ever-more-sophisticated decision-support packages, that still holds true.

The fourth finding could only emerge out of that actual experience. This was the way fires over time affected one another. Particularly in landscapes prone to stand-replacing fires, each big burn acted to buffer and check the next. The SBW began to fill out with burns, and even to reburn. Physics might appear to underwrite fire behavior, but it was history that underwrote physics. Fire regimes were historical constructions. They emerged from decades and centuries of fires burning, scorching, reburning, scouring, and just plain skipping around; of fires that burned in patches, and of big burns that burned out patches; of fires that blew themselves out when they struck recent burns; of fires riding over the troughs and crests of droughts and heavy snow years and winds capable of blowing down whole woods. Such patterns could not be reconstructed from first principles over a season or two, or even a decade or two. They required a long commitment. Probably they would demand centuries.

The Selway-Bitterroot Wilderness was an ideal site for such an experiment. Its remoteness and, for commerce, its emptiness had kept damages to a minimum; it had resisted all-out suppression as much as Florida had.

Fire control did not begin to gain traction until after the Civilian Conservation Corps arrived, and then not until the postwar era let fire suppression take to the air. Probably serious suppression involved no more than two decades before the White Cap Project began to exchange tolerated natural fire for suppressed wildfire. The disruption of fire regimes was far less than for montane forests or parkland forests that had burned every 2 to 5 years and had excluded fire for perhaps 80 to 100 years, or missed 15 to 20 fire returns. Still, for a place that had been declared the test site for the 10 a.m. policy, a natural fire program was a bold decision. The progeny of the White Cap Project had chased suppression to its lair.

There was plenty yet to do in the SBW. Much of the landscape had still not burned even once, while a genuine regime required reburns. But the dynamic was unmistakable. Every fire suppressed made it more difficult to suppress the next fire. Every fire allowed to burn made it easier to allow future fires. There was no neutral position. You let fire do what it needed to do, or you broke the rhythms once more and added to the ecological debt burden. Moreover, the best way to simplify the process was to continue to expand outward over all the wilderness and as much quasi-wilderness as possible. Restoration needed room and time. The SBW still lagged. There persisted a misbegotten legacy to overcome; the forest borders, bristling with new houses, were becoming more intolerant of fire; as personnel cycled through the many districts (four forests and two regions), newcomers had to be initiated into the regime when they passed through the portal. Not least, suppression still lingered as a rootstock, ready to send out shoots from below the graft of natural fire management. Still, the burning was accelerating.

The SBW absorbed lightning-kindled fires with little expense and scant risk unless one discounts, which one can't, the sweaty palms of higher administrators. Yet the contrast was undeniable. To one side of the Bitterroots the future pointed to more ravenous fires, ballistic costs, and a vicious spiral of self-reinforcing suppression. To the other, the future pointed to ecologically enhancing burns, self-limiting costs, and a virtuous spiral of self-contained fires. The present, at always, was a muddle, and it was often hard to determine at any moment what cycle was spinning.

On May 31, 2012, Deputy Chief James E. Hubbard dispatched a memo to regional foresters, station directors, area directors, the International Institute of Tropical Forestry director, deputy chiefs, and Washington Office directors laying out guidelines for the coming season. The text opened with a boilerplate reaffirmation that fire management was "central to meeting the Forest Service mission" and that, among the many tasks of that mission, the protection of life and property was primary.

Then came the bad news. The season promised to be a tough one, suppression costs were forecast to exceed the 10-year average (and allocated budget), money to finance the firefight was tight and would have to come from other Forest Service programs, which would dampen the chances to do other tasks and alienate various and oft-vocal publics. The two poles that had torn wildland fire over the past 40 years—the wild and the urban—threatened to rift rather than meld. Congress wanted houses protected more than it wanted wilderness burned, and no one wanted anyone to die for either cause. Certainly some of the worst wildfires of the new era had originated as prescribed burns or wildland fires given a long leash that then went feral. Rumors circulated, too, that if the Forest Service busted its fire budget this year, responsibility for fire would be handed over to FEMA, a move that would destroy every vestige of good achieved during 50 years of the American fire revolution. Accordingly, a Risk Decision Framework was specified to guide decisions. It directed that any choices that might involve allowing fires to free-burn would be taken from the hands of local officials and put into those of the regional foresters. The bottom line was—the bottom line. To protect costs aggressive initial attack, everywhere, would be the first response.

The memo came with a series of talking points that assured readers that the decision was only temporary, a reaction to the unholy collusion between a worsening climate and a weakening federal budget, and that its authors acknowledged, and emphasized "that such an approach is not sustainable over the long-run." Critics had long recognized that firefights were like a declaration of martial law, able to put down an ecological insurrection or the occasional riot but not a means by which to govern the landscape. The historically minded might note that just such economic logic had guided the 10 a.m. policy, which Chief Forester Gus Silcox had announced as an "experiment on a continental scale." The big fires had run up the big costs, and the simplest way to prevent big fires was to kill (and overkill) all fires while they were small. Now, it seemed, the old experiment was being rerun. Fire and line officers wondered just what, in practice, the new orders meant. The simplest solution was surely to suppress everything. No one would be faulted for aggressive initial attack.

The sympathetically minded might note that, perhaps, this really was a temporary blip. But such mandates, like temporary taxes, tend to persist. The memo sent a clear message that the safest approach—safest for careers if not for life, property, and land—was to suppress every start. The sympathetic might also argue that the guidelines could be interpreted as the final nail in the coffin of the old prescribed natural/wildland fire use/resource benefits fire category, that all fires, as wildfires, could be managed in ways to promote ecological enhancements however they were labeled. But that is not what the interlinear text said. It said "aggressive initial attack" except where unsafe or imprudent to do so. An ecological insurgency, however, can be suppressed by sending in a fire constabulary for a while, but eventually it will boil over and consume the regime. Was the fire scene like the financial meltdown of 2008, requiring a temporary if unpalatable bailout? Or was it like an Arab Spring, in which the longer repression remained the more likely an outright civil war would result?

There was, in July 2012, no way to say. What a century of fire suppression in the Northern Rockies had taught, however, was that the big fire years led to the big burns that did the biological work required and that helped to contain the blowups of the future. Allowing fires except in those critical years was a charade. Worse, what the Predictive Services emanating from fire sciences was forecasting was not just an upcoming season of exceptional demands, but an upheaval of climate that might make the temporary blip into a new normal. Suppression could, at most, create only an illusion of security, and in any event would soon blow through its budgets.

So the edict would go to the court of nature for review. There are ample examples in which, however desperate for fire restoration, officials decided that this particular fire at this time and place—amid an extreme season, next to major highways or housing developments—was not the one to loose. It might be that the Forest Service was making an equivalent choice for an entire season. There are equally many examples, from life as well as the woods, where such decisions are simply a false caution that is indistinguishable from a failure of nerve or outright bureaucratic cowardice. First establish order, then reform is the cry of counterrevolutionaries everywhere. The effort to impose that order becomes itself a process of further disordering that typically concludes with open revolt.

———

Dave Campbell said nothing about the national directives. The Selway-Bitterroot Wilderness has been so long in evolving that it was almost unthinkable to imagine a return to the ancien régime. Yet it was also clear that its power as an exemplar might be held to the scrawled borders of legal wilderness and the peculiar standing of the SBW in the national fire saga. It was possible that no tablets of instructions—what is ultimately an expressed ethic—might emerge from its burning bush. Yet for two hours it had been enough to savor the hardwired testimony of the Selway.

As the Cessna touched down on the tarmac at Hamilton, Bob Mutch observed, "We've come full circle." It remains to be seen which circle will best express his words.

FIRE BY PARALLAX

The Flathead Reservation

I N 1842 FATHER PIERRE-JEAN DE SMET traveled by invitation to the Coeur d'Alene tribe to begin missionizing. Later that year he was joined by another Jesuit priest, Father Nicolas Point, and a lay brother, Charles Huet. Originally from Belgium (as was Point), Smet had emigrated to the United States in 1821, ran a school for American Indians in Missouri, established the Potawatomi mission at Council Bluffs, and in 1840 had traveled to the Northern Rockies at the request of the Flatheads, where he was greeted enthusiastically and traveled widely around the region. The excursion to the Coeur d'Alene, another Salish-speaking people, was an echo of that momentous event.

While there Smet wrote accounts of what he found, and Point sketched and painted people and scenes—a kind of Jesuit George Catlin. Among that record is a painting of a fire hunt in which flames are driving deer into the lake where hunters in canoes can easily kill them. This was not an uncommon technique; there are reports of indigenes similarly driving deer into the tidewater of Virginia, and Fenimore Cooper describes hunting deer from canoes in *The Pioneers*. But the painting records something else, for it depicts the same flames as savaging the hunters' village. Almost certainly the immolating encampment is something Point inserted into whatever he and Smet witnessed.[1]

In a few strokes, however, Point's painting captured an ethnographic practice, an ethnocentric perspective, and the seemingly exclusive interpretation that could overlay them as two peoples viewed the same scene with different eyes. For the Belgium-born Jesuits burning the landscape was the same as burning one's house, for it destroyed habitat and was an expression of what, as much as with spiritual beliefs, they had come to reform. For the Coeur d'Alenes landscape fire was part of what made that land habitable. They burned seasonally for hunting, for camas and berry production, for pasture, and for the protection of villages. For the Jesuits landscape fire was, in a sense, hellfire on Earth and a symbol of the perdition they had come to replace. For the Coeur d'Alenes—and the other tribes that inhabited the Northern Rockies—fire was a gift of the Coyote and its possession was much of what made them human.

That well-intentioned and mutual misunderstanding of what fire meant would continue. Point's painting is a visual metaphor of what was to follow with regard to people, land, and fire, and the way each side could read the other wrongly. In reality, two images coexisted, not so much side by side as overlaid, as though viewed through a stereopticon. They might appear to the casual observer as one, but they were less a synthesis than a holographic card that assumed one image or another as the card tilted. They remained separate, however much blurred by visible emblems of acculturation such as religious conversion, adoption of farming, or going to school in brick buildings.

At the time of the Smet mission, Salish speakers claimed most of the Northern Rockies and spilled west and north. Three tribes became especially significant: the Flatheads in the Bitterroot Valley and the Pend d'Oreilles and the Kootenais, a non-Salish group, both farther north. Remarkably, observing the thickening swarm of white transients and settlers, the Flatheads requested the Black Robes to visit them as they sought to learn about the newcomers and their powers. The missions commenced in 1841. Acculturation was seen by many as a means of resisting, or at least allowing the option of a choice to retain power over their lives not possible by overt fighting. The Nez Perce and Blackfeet, west and east, showed the alternatives.[2]

The larger trends were determined by powers far removed from the mountains. In 1846 the United States and Britain divided the collectively held Oregon Territory along the 49th parallel, accelerating the political assimilation of the Northern Rockies. Subsequent history hinged on the Hellgate Council (near Missoula), convened in 1854 by Governor Isaac Stevens to present the terms of a treaty that would establish a common reservation, formally identify tribes and chiefs, and create terms of engagement. Stevens insisted on a single reservation for the Flatheads, Pend d'Oreilles, and Kootenais, all of whom he treated as one tribe. The reservation's location would be determined in the future. Other peoples might also be housed there.

Each side viewed the incident and its outcomes differently. Discussions were conducted through interpreters, refracting through terms and concepts peculiar to each people. "Very likely," as historian John Fahey observed, "neither the white nor the Indian negotiators understood one another well." In the end, Stevens got the Hellgate Treaty, signed in 1855, on the terms he wanted, and the newly confederated tribal nation got a chief they didn't recognize, a collective polity they didn't seek, and a reservation they couldn't yet locate. The Flatheads, in particular, worried that they might lose their Bitterroot Valley homeland, which proved correct.[3]

What followed was viewed by white observers as a story of successful acculturation and by the tribes as one of alienation. Through the usual political pathologies and legal shenanigans, the confederated Flathead nation lost most of its ancestral homelands. While their new reservation in the Flathead Valley was large by the standards of small-settler homesteading, it was tiny for a people who were accustomed to seasonally migrating over the mountains and to the plains for sustenance. Even that reserved land began to flake off. In 1882 the Great Northern Railway bought, against tribal wishes, a 53-mile right-of-way that cleaved through the collective allotment. Forest reserves spalled off some two-thirds of the reservation: the Flathead and Bitterroot National Forests (1897); the Missoula, Kootenai, and Lolo (1906); the Cabinet (1907); and the Blackfoot (1908). The Dawes Act of 1887 fractured the tribal commons into 2,460 parcels of 80 and 160 acres, along with a dozen town sites, a survey not completed until 1909. That year a National Bison Range was established on 18,500 acres of reservation land. Meanwhile a 1907 agreement allowed the U.S. Forest Service to oversee the tribe's still-retained wooded lands.

But population and identity were lost as well. Smallpox and other introduced diseases swept away significant fractions. Treaties reconstituted the extant population by combining tribes, and adding peoples even outside the Salish linguistic orbit. At one point, despite tribal protests, Canadian Crees and métis were folded into the demographic mix. Intermarriage was common, not only among the subtribes but with the surrounding white communities. In 1900 the reservation held a population of 1,734 persons, half of whom were classified as mixed bloods (mostly French Canadian). With the 1934 Indian Reorganization Act, the Confederated Salish and Kootenai Tribe replaced the misnomered Flatheads. Like its reservation, a tribal identity was both displaced and imposed.

The losses reached a climax in 1910. The tribe lost further control over its lands when unclaimed homestead lots were opened to the public, which hollowed out the arable middle of the reservation (of 1.3 million acres, some 700,000 remained tribal). It lost further local control over its forests as the agreement with the Forest Service was terminated, and a Forestry Branch, authorized the year before, was funded and established within the Office of Indian Affairs. And it lost control over fire, as the Great Fires—"in nearly every point"—swept over the reservation. The Flathead Agency was wholly unprepared for flames of this magnitude. It hired gangs of workers, and when they proved ineffective, it twice requested assistance from the U.S. Army, which dispatched two companies from Washington and North Dakota and two from Fort Missoula. The fires gutted the embryonic forestry plans; some 60,000 acres of mostly young timber burned off, and the salvage logging of the killed mature timber was undercut by a market flooded with dead forests throughout the region.[4]

Even amid a tribal history replete with symbolic inflections, the year stands out.[5]

———

Like a campfire, so common and expected that it is as unnoticed as it is essential, landscape burning had always been a part of the annual circuit of the Salish economy. Every season had its fire: the winter buffalo hunt, the spring harvest of camas and bitterroot, the summer hunting and fishing, the fall gathering of berries and roots. The Point painting of a fire hunt was inaccurate on several counts, but especially egregious was its alignment of where people lived and where they burned. Rather, they resided in one place, with campfires; they burned in other places, with broadcast and spot fires.[6]

As the tribes contracted, pressed by enemies on the Columbian Plains and the Great Plains, and by white intruders, their fires no longer remained in the old mosaic. To live the old way was to burn the old way, but the newcomers sought a novel pattern of land ownership that left fires to the mountain reserves. During a hunt between 1874 and 1876 two Indians were shot by settlers for setting the prairie afire. Even when escorted by troops on their circuit, they slipped their leash long enough to kindle traditional grass and camas prairie fires and prompt complaints by settlers. The 1910 fires began as spring burns by berry pickers, later supplemented by the promiscuous kindlings of white prospectors. Probably, as J. W. Powell had documented along the Utah mountains, settlement pressures were forcing old fire practices to push into new settings. The mixing of tribes had also mixed fire habits for which a common land did not provide a common policy. The sequestering of the indigenous burners went a long way to sequestering fire. But railroads added new ignitions. Then, as wooden towns arose, fires even swept through stores and schools.[7]

The new order of fire began with forestry, and it dates, as so much of the Northern Rockies fire scene does, from the Big Blowup. The fires coincided with the creation of a Forestry Branch within the Office of Indian Affairs, which put fire protection on the national agenda and confirmed forestry as the received oracle for all matters pertaining to fire. The Flathead Agency began hiring seasonal forestry guards who could serve as an on-site fire crew. By the late 1920s some 15 to 26 guards served each summer. Improved roads and mechanical transport quickened initial attack; a speeder patrolled the Northern Pacific Railway lines; and an annual campaign of fire prevention was inaugurated. Cooperative agreements with the U.S. Forest Service helped align practices on the reservation with those on the national forests, which also provided the only fire lookouts.

The big change came with the New Deal. The Indian Reorganization Act helped modernize the Bureau of Indian Affairs, while the Civilian Conservation Corps (CCC) almost overnight created the infrastructure needed for fire control in the Mission Range, Salish Mountains, and Hog Heaven Range. In 1931 the Flathead reservation had a solitary lookout on Saddle Mountain. By 1942, when the CCC ceased, it had seven, all joined by telephone lines and a radio station. More roads divvied up the unbroken countryside, providing better access. The CCC furnished fire crews. Moreover, the Corps served as interagency sinews to bind fire protection on the reservation to the neighboring forests. Cooperative fire protection expanded from mutual aid agreements with the Forest Service to include contracts with the State of Montana and private lands under the Northern Montana Forestry Association. The Flathead Agency (after 1934, the Confederated Salish-Kootenai Tribe [CSKT]) created a facsimile of Forest Service programs. It did so, however, without a comparable funding mechanism. Its annual report for 1959 noted that, while they appeared similar, the CSKT's fire program—personnel, equipment, maintenance of trails and repair of towers—was rapidly falling behind its federal neighbors. Everything needed updating. The stereopticon effect that had characterized fire understanding since Nicolas Point's allegorical painting had assumed a modern, mechanized update.

Fifty years after the Great Fires, the reservation suffered a bout of big burns that began in early July 1960 and, bolstered by dry lightning storms, had reached over 100 ignitions by July 20. The fires blew away the increasingly flimsy tissue of fire protection erected over the past decades. The CSKT doubled down. It signed a cooperative smokejumper program with the Aerial Fire Depot in Missoula; later created a helitack crew; trained fire crews on call-up out of forest workers; and hired professionals out of the Forest Service. It started a prescribed burning program to handle the fast-amassing slash from accelerated logging operations. It joined the USFS in constructing a lookout on Baldy Mountain. Over the next decade the program rebuilt. It hired fire officers from the Forest Service. It looked like fire programs elsewhere in the region. It became a part of the Missoula matrix, although like the Flatheads it had relocated from the Bitterroot Valley proper northward.

By the late 1970s the image of the Flathead Reservation resembled the scenes around it, save for poorer funding. Tilting the image a bit, however, revealed another image beneath. Like Point's painting, reality was an unstable composite made from an overlay. Already events were twisting in ways to bring the two images into sharper parallax.

What happened was the fire revolution with its ultimate ambition to restore free-burning flame. For the national forests and parks of the Northern Rockies, this meant allowing natural fires to burn, or creating the conditions that would permit such fires, or substituting prescribed fires where untrammeled burns could not be tolerated. For the Flathead Reservation, however, it meant restoring something of the landscapes and fire practices that had prevailed in pre-Columbian times. This was a deeper reform made possible by a parallel revolution in Native American governance. The two restorations overlay each other.[8]

In the 1950s the long trend to assimilate had led to a policy of outright termination—the cultural equivalent to fire suppression. The resulting reaction occurred in lockstep with the fire revolution. In 1961 a pan-Indian conference convened at the University of Chicago and led in 1962 to a "Declaration of Indian Purpose" presented to President Kennedy. At the same time the National Indian Youth Council organized as an advocacy group for a new generation. A series of reform legislation and court decisions began addressing education, health, tribal courts, access to natural resources (notably, fishing and water rights)—these culminated as a protest movement in the American Indian Movement and as a legal regime in the American Indian Civil Rights Act (1968) and then the Indian Self-Determination and Education Act (1975). An American Indian Policy Review Commission oversaw a panoramic survey, which led to a report in 1977. All this set in motion a series of restorations—of tribal identity, of religious sites, of historic sites, of control over natural resources and even the return of some ceded lands. What happened in fire, as a national program moved from a hegemonic agency with a single policy to a pluralism of lands and practices, had a parallel in American Indian history. At places like the Flathead Reservation the two movements braided together.[9]

Throughout the 1970s the CSKT acquired more control over its forestry and fire programs, even as it emulated the planning guidelines and practices of the other federal agencies and commissioned a detailed history of its forestry program. In principle, the BIA's charge to hold the land in trust obligated it to emulate the best programs, which also meant funding them at comparable levels. In practice, it had tended to add encumbering layers of bureaucracy while never financing programs adequately. What had been a separate but equal doctrine had inevitably proved neither separate nor equal. In 1992 the CSKT accelerated its move toward self-government, including its management of woods, waters, and lands generally. It sought working landscapes; production forestry continued; and revenue helped finance other programs. But it also wanted special areas, the equivalent of wilderness. A primitive area had been proclaimed in 1979; in 1982 it became the Mission Mountains Tribal Wilderness—the first tribally mandated wilderness nationally. This was supplemented by three special use areas; the South Fork of the Jocko, Lozeau, and Chief Cliff. Throughout, the proclaimed norm was a restoration to more pre-Columbian conditions or at least ancestral attitudes adapted to more modern times. That included fire.

A complex process of planning commenced that resulted in a kind of alternate version of multiple-use land. An interim forest management plan emerged in 1997, an environmental impact statement in 1999, and a new forest management plan in 2000, coincident with the National Fire Plan, with a dedicated fire management plan in 2007. Mostly prescribed fire remained in logging slash, but some underburning was spreading, particularly in ponderosa pine savannas, and there was a willingness to grant fires in the wilderness some play, especially since much of the adjacent lands were Forest Service wilderness. Yet tribal wilderness was not the same as wilderness on federal lands, and an acceptance of fire did not originate solely from the same wellsprings, so part of the commitment to restore something of the old fire regimes was to recover the old lore that had underwritten human occupation of the land. The upshot was a remarkable summary of legends, techniques, and regimes coded into an interactive DVD and website, *Fire on the Land: a Tribal Perspective*, released in 2004. Where the national forests and parks looked to natural science for a foundation to practice, believing that the best landscape was one that was as wholly "natural" as possible, the CSKT accepted a former cultural landscape as its ideal and turned toward traditional ecological knowledge for guidance. Both perspectives looked to the past, but one peered into a past before humans and the other into that past before Europeans. One appealed to wilderness and natural science, both presumed transcendent and above culture. The other appealed to culture, from which concepts like wilderness and practices like science derive and through which they must be nourished.

Today, the Flathead Reservation has a fire program that looks much like those around it. It fights fire on its timberlands and burns slash. It prescribe burns select tiles in its landscape mosaic. It encourages, on a case-by-case basis, some wilderness fire, and may intervene to put more back in if needed. It suppresses fire under contract for some state and rural lands. It supplies 20-person crews for work off the reservation. It meets comparable standards for training, operates identical equipment, attends the same conferences, and serves on shared National Wildfire Coordinating Group panels. For all intents and purposes, it is interchangeable with other major players within the national system and regional complex.[10]

And yet it has fashioned an alternate vision, an overlay in which the commonalities, on casual inspection, have blurred the distinctions. While the CSKT fire program is encumbered by layers of bureaucracy and oversight and legal filters not true of its allied fire agencies, it is also spared some of the public and political scrutiny that constrain the others. It has freedoms to act that they lack. It has what it calls wilderness, but is not subject to the Wilderness Act. It can have its own airshed plan, and manages smoke accordingly. Even when it engages in endeavors similar to those around it, those practices are refracted through a different perspectival prism. It tells a counternarrative in parallax.

———

In the modern era the Confederated Salish Kootenai Tribe's most famous member has surely been D'Arcy McNickle. His biography encapsulates perfectly both the blending of peoples and the incomprehensions that divide them. He had an Irish father and a Cree métis mother; he could live in both tribal and white worlds; he could seem to whites both acculturated and progressive, and to Indians a defender of rights and tradition. He became a major figure in the movement for national reform. And he wrote three novels that exactly span those times and illustrate brilliantly how two people could, even with the best of intentions, read a scene in diametrically opposite ways.

The first, *The Surrounded*, was published in 1936, two years after the Indian Reorganization Act. The last, *Wind from an Enemy Sky*, appeared in 1978, posthumously, three years after the Indian Self-Determination and Education Act. The plot of *Wind* turns on the construction of a dam. To whites it is a marvel of applied science, a taming of natural processes to better economic purposes. To Bull, the Indian protagonist, the act is insane: "They can't stop water. Water just swallows everything and waits for more. That's the way with water." Worse, the dam buries a sacred site. Each side holds to its culture and the perspective it encourages, while their interaction unfolds into tragedy. What McNickle wrote about water might equally apply to fire, except the tragedy is borne by the land, not persons.

Throughout, the two cultures have not only viewed fire differently, but have incommensurate understandings about how to view a firescape at all. Today, it is possible to imagine a remake of Nicolas Point's painting in which crews light the woods to promote ecological benefits while deer and bison forage on the refreshed browse. The prevailing perspective would explain that scene as a reinstatement to a better, more natural world in which fire has returned to work its biotic alchemy, with the implication that keeping the fire and removing the fire-starters would improve the scene even further. The firescape should be viewed through the moral prism of wilderness and the analytical prism of science. Another perspective would parse that image as a restoration to a former world in which people had coexisted with those woods and grasses and creatures and did the task uniquely assigned to them: they burned. In this rendering, to remove the people would unhinge the fire and unbalance the prelapsarian order—the Fall in this case resulting from the encounter with Europe. It is a firescape encompassed by culture, however decorated along its margins, like an illuminated manuscript, by the machinery and clothing of modern technology.

That the two visions can coexist is possible because places like the Flathead Reservation survive.

THE EMBERS WILL FIND A WAY

The mind is not a vessel to be filled but a fire to be kindled.
—PLUTARCH

B EGIN WITH THE FUNDAMENTALS, which is where a fire science that wants to think of itself as applied physics likes to begin. The first question is how a fire gets started and the second how it spreads. Know that and the rest follows. The same might be said about fire scientists.

In 1950 when Jack Cohen was born, Tucson had a population of 45,000, and the family homestead was in any event closer to Mission San Xavier del Bac. The landscape was open high desert split by the riparian cleft of the still-flowing Santa Cruz River. When he left 18 years later, having graduated from high school at 16 and having spent two years at the University of Arizona, the city had swollen to 263,000 and in its search for water had pushed wells so deep that the Santa Cruz had become a ditch. It was an object lesson about sprawl and its environmental disruptions, but Jack was not one to look back. By then he had become the equivalent of a firebrand lofted into the prevailing winds, looking for a suitable place to alight.[1]

What gives that simile some verisimilitude is that he spent his youth starting fires. He was the one to kindle whatever fire someone needed set. He burnt off the lawn in spring. He burned leaves in a burn barrel. He ignited the barbecue. He shot firearms, which in an age before ear protection ruined his hearing. If someone needed a cigarette lit—this was before the surgeon general plastered warnings on packages—Jack volunteered to make the flame. The family had a summer cabin in the Santa Rita Mountains amid the Coronado National Forest, a place readily sparked by summer lightning storms, and prone to fires, although aggressive firefighting held nature's electrostatic match to sparks and snags.

Such were the basic facts, but like principles they needed some way to combine to put fire on the ground.

His youthful fireworks were all play; there was no malice and no one got hurt. But it was not magic, because Jack Cohen was already an inveterate scientist who wanted tests, numbers, and proofs. For years he struggled, however, to find the right science. In high school he turned to chemistry, and majored in it at the University of Arizona; and in important ways he never abandoned it. Meanwhile his interest in outdoor recreation grew. But nothing gelled. He did well in subjects he liked, and poorly in those he didn't. He took a year off, worked on drill rigs in New Mexico, and learned some practical geology; the rigs destroyed whatever remained of his hearing. He realized what he did not want to do with his life.

In 1971 he decided he wanted to be a forester and transferred to the University of Montana. He soon became disillusioned with forestry, took the minimum courses, and loaded up on botany, physics, and math. The critical moment of his Montana sojourn came in the summers of 1972 and 1973, when he worked as an assistant to UM professor Bob Steele, who had contracts for prescribed fire research. Jack had to learn everything: not just the science, but how to build line, work with hand tools, and set fires. That's what did it: direct contact with flame. For someone who thinks visually and tactilely, not verbally, it was a baptism by immersion that merged idea with act. Jack Cohen found himself touched by fire in ways that jostled together the various parts of his unsettled life into something resembling a coherent pattern.

He went to Colorado State for a master's degree, although, still bored with forestry, he enrolled in the atmospheric sciences program. But he craved a fire connection both for hand and mind. He found one by working his first summer on the Pike-San Isabel hotshot crew, which expanded his practical understanding of fire operations but left him again twitching with intellectual impatience. He resolved that issue by switching departments to biometeorology and bioclimatology, which led him to contacts with Forest Service researchers slowly assembling a National Fire Danger Rating System (NFDRS), a grandly synthetic program that brought everything known to bear on when fires might be expected to start and how vigorously they might spread. His 1976 master's thesis processed the fire climatology of the Colorado mountains.

What had been mostly playful now turned professional. He was immediately hired by the Missoula lab to help develop the NFDRS, which was released in 1978. And he switched from cutting fireline seasonally to analyzing fire behavior for going fires, graduating from the third fire behavior officer class. The NFDRS was about doing fundamental science but around a conceptual core that mattered to the field, and fire behavior officers took the best of that understanding and spoke to the overhead teams on big fires. The boy who grew up eager to light fire and easily bored had found a suitable fuelbed to alight on.

Fire was his field—that was confirmed. But fire burned in many settings, and Jack found himself blown from one to another. For the next 20 years he cycled through topics and Forest Service fire labs, at one point resigning altogether for 18 months. Throughout, there were two constants—his curiosity about ignition and the need for direct contact—to connect what was known with what was done in the field.

In 1979, a year after the NFDRS was released, he transferred to the Riverside lab in Southern California. Here he met new topics like Santa Ana winds, Mediterranean climates, and chaparral, but they were also variants of old ones; the fundamentals were the same. What was most novel was the urgency of unraveling how live fuels burned and how fire actually behaved in the fractal frontier between houses and wildlands. The need for experimental fires led him to certify as a Lighting Boss on the San Bernardino National Forest. He reconnected with fire teams as the South Zone fire behavior officer. The need to put knowledge—the right knowledge—on the line persisted.

But fire had a way of crossing lines. It crossed from wildland to urban fringe, which also breached methods of fire suppression. The fabled foehn winds over the Transverse Range flung embers for miles, which mocked fuelbreaks and firelines. So, too, flames on the mountains broke through the parameters of models developed in the smooth beds of laboratory wind tunnels. Conducting field burns and doing tours with overhead teams mixed the administrative borders between research and operations. When in an internal report he criticized the prevailing Rothermel model of fire behavior, now the nuclear core of the NFDRS prediction program, as unable to forecast the kind of fire behavior he was witnessing in the field and as propagating only "an illusion of understanding," he was reprimanded.

By the mid-1980s fire research was on the ropes. The Reagan administration sought to reallocate government science from civilian to military purposes and, buoyed by a bout of wet years in the West, chose to believe the fire problem was solved. Funding was gutted, and since the Forest Service had a virtual monopoly over fire science, that sent the entire field reeling. In many respects, the agency never recovered; it could sweep up the pieces but never quite reassemble them. The Forest Service tried to salvage what it could with a major redefinition of "the fire problem" and a massive reorganization announced in July 1986 that tended to anger and unhinge its staff. The reorganization proposed to send Jack to the Macon lab. "Uncompromising" family concerns, however, led Jack to resign and remain in Riverside. For 18 months he worked as a research engineer for Dunham-Bush Inc., a company that manufactured heat exchangers.

In August 1988 he returned to the Riverside lab as a full-time temporary employee, and was fully reinstated a year later as he completed his transfer to the Macon lab. His assignment was to help adapt the NFDRS to the Southeast. But by the time he arrived the applications mission was being superseded by a new quest for fundamentals. The American fire scene was itself undergoing a continental-scale transfer of themes and attention as the ignition of houses became as critical as the kindling of wilderness. The wildland side had an NFDRS to forecast ignition and spread probabilities. The urban side needed an equivalent.

It was an inauspicious time for a research initiative, but desperate to find a topic compelling to the administration, Forest Fire and Atmospheric Sciences Research identified what was awkwardly termed the wildland-urban interface (WUI). This was a place where an implacable urban sprawl met irrepressible wildland fires and hence collided with public safety and politics. Those in Forest Fire and Atmospheric Science Research hoped that it might give handholds for a research organization otherwise stranded on a cliff face.

For 25 years, since the Bel Air-Brentwood conflagration had swept over the Hollywood Hills, the problem had spread like blister rust, infecting the American landscape but without much visibility until it occasionally broke out in epidemics. The worst were in California. In 1985, however, the California plague went bicoastal. On the West Coast fires ripped into Baldwin Hills, Los Gatos, and Ojai; Jack self-deployed to the first and went to the others on assignment. On the East Coast a fire savaged the community of Palm Coast, Florida.

Suddenly, what had been a peculiarity of California exceptionalism looked like the next new thing in fire protection. The wildland-urban interface fire—"a dumb name," one critic noted, "for a dumb problem" because technical solutions existed—had become a national issue. The next year the National Fire Protection Association (NFPA), alert to the emerging crisis, partnered with the Forest Service to sponsor a conference, Wildfire Strikes Home, that became the basis for a program termed Firewise. NFPA took the lead. The hope that the topic might relaunch fire research faltered until the NWCG accepted a wildland-urban interface initiative in 1995. Jack Cohen served as Forest Service advisor.

At first blush it was a topic that seemed remote from wildland fire. But like the houses that crowded against wildland boundaries, its themes shared a common border in combustion science. The critical questions were identical: how did structures ignite and how did the subsequent fires spread? Better, it was a subject around which various Forest Service labs could rally. By 1989 the contours of a research program had become visible through the smoke. The pivot was the Forest Products Lab in Madison, Wisconsin, which took an interest because wood frame houses were the principal fuel. It wanted to understand ignition properties of wood as it did the strength and durability of construction timber. Initially it sought to create an expert system to identify the critical issues for research but soon discovered that, as Jack forcefully declaimed, there were no experts. Instead, understanding would have to come from fieldwork amid incinerated communities and from laboratory research into the fundamentals.

The Macon lab, excelling in combustion basics since the days of Wallace Fons and George Byram, took on topics in heat transfer and ignition. The Riverside lab had the problem in its backyard—an uncontrolled field experiment—ready for empirical analysis. The Missoula lab, having sponsored the prevailing model of fire spread, should have been as vital to the WUI issue as it was to the NFDRS, but fire behavior in suburbs was not a topic that could be tested in its wind tunnels, and doubts among the knowledgeable grew about the validity of its core model, which assumed that fire spread by continuously radiating fresh fuel from an advancing flaming front. Real fires seemed to burn in patches and belched in billowing plumes; there was little continuous about Rocky Mountain forests or subdivisions sprouting on a once rural countryside; and some blasted communities had their houses burned to concrete stubble while surrounding trees remained green. Still, however dense the institutional combustibles, the WUI as a research topic did not self-combust. It needed a firebrand. It found one in Jack Cohen.

The disruptions from sprawl, of unthinking development destroying its own habitat, he had known from his childhood. But throughout his career the flaming fringe had followed him like a peripheral glow. He recalled as an undergraduate watching a documentary on the Bel Air-Brentwood fire, *Design for Disaster*. Then he saw equivalents burn without the filter of editors. As a new hire at the Missoula lab, he had observed, as a volunteer fireman on the ground, the Pattee Canyon fire leap through Missoula exurbs without, as he recollected, much of a flaming front. A crown fire had dissolved into a firebrand blizzard. As a fire behavior officer in Southern California, he had considered houses as another cache of combustibles, and again after the 1980 Panorama fire he had reconstructed the dynamics of ignition as wooden roofs broke into flame half a mile beyond the fire's perimeter. Such oddities of ignition had been, as it were, beyond the borders of his professional concern. Now they began to gather into a coherent agenda of questions.

It fell to him to develop a prototype model. What he discovered did not fit existing concepts but did follow his understanding of how science worked, by swarms of ideas searching for tests. Ninety percent of science, he decided, was asking the right question. He learned that windows fractured from radiated heat before exterior walls ignited. He found that fires did not rush across fringe suburbs like a tsunami of flame, but blew over and through them in blizzards of sparks. Downstream of the plume or flaming front there were thousands of sparks, and even a mile away there might be scores. With so many swirling firebrands in an ember storm, it was unlikely any point of vulnerability would escape. Like the rain that falls on the just and unjust alike, the embers fell on everything. If any way were possible to kindle a propagating fire—on a wood stairway, in a collar of pine needles around a yellow pine, amid a town dump—the swarming embers would find it. By 1993 he had consolidated his insights into a prototype, the structure ignition assessment model, that shifted emphasis away from the flaming wildlands and onto brand-receptive houses; that also shifted the power of control from wildland agencies to what was in fact, if not acknowledged, an urban fire scene.

The insight was so counterintuitive that it seemed to defy common sense. As he came to dissect it, the whole WUI was a trompe l'oeil. The problem was not about the soaring flames but the swarming embers, not the wildlands but the urbanizing fringe, not the flammable woods but the combustible houses. To see it required a gestalt-like switch: the silhouette of a tree morphed into a mountain. So, too, in the mechanics of this new fire scene the lowly (or obscure) trumped the obvious (or celebrated and iconic); and that could be said equally about fire models and researchers. But then, to Jack's mind, it had always been the role of fundamental science to overturn common lore with positive knowledge. The great ideas would always blow across the lines scratched in the duff by institutions.

His discoveries did not immediately fit into fire protection schemes that were after all, and had been for most of a century, directed at halting the spreading flames. But neither did the agenda of the Macon lab fit perceived national needs. In 1995 the Forest Service, still consolidating its fire research programs, like a body in shock pooling blood away from peripheral limbs and into vital organs, ceded the building back to Georgia and shuttered the lab. Jack returned to Missoula. The structure ignition research program returned with him.

Throughout, he visited and investigated in detail the major WUI conflagrations; Baldwin Hills, Painted Cave, Los Alamos, Rodeo-Chediski,

Hayman, Aspen, Angora, Grass Valley, Myrtle Beach, and Fourmile Canyon. Initially he had been invited to accompany teams fielded by the NFPA. Then he began self-deploying: he needed to see firsthand where the flame made contact. He became, in effect, an unofficial fire behavior officer for the WUI. More and more, the story of conflagrations proved to be a tale of tiny kindlings that grew monstrous. Then the International Crown Fire Modeling Experiment offered an opportunity to field-test theories of radiant heat transfer and house ignition. The trials confirmed that radiant heat—so obviously seen and felt—was a minor contributor. What mattered to exterior walls was flame contact. And what mattered to the textured structures that made up a house were niches to accept embers. As a colleague phrased Jack's argument, the embers will always find a way.

In 1999 Jack presented a paper on his findings to the International Conference on Fire Economics and sparked an uproar of controversy when environmentalists realized that the growing emphasis on whole-sale fuels treatments on public wildlands, proposed as a means to protect exurbs, misdirected attention away from what Jack had termed the "home ignition zone," located on private property. His research challenged the scientific testimony behind what, following the epochal 2000 fire season, became the National Fire Plan. He felt he had become persona non grata for appearing to question the evidence behind evolving presuppression and fuels programs.

Still, in 2002—as record fires romped across Arizona, Colorado, and Oregon and crashed into classic sprawl—he released a popular video produced by the NFPA-Firewise program (*Wildfire: Preventing Home Ignitions*) that distilled what he had learned. The core problem, he continued to insist, was protecting the structure at risk, not suppressing the fire in the wildland. Wholesale treatments of woods, expensive and dangerous mustering of crews and engines—these could be useful but only if they did not misdirect attention from the home ignition zone. It was the point of contact that mattered. The rest was a distraction, and he worried that it might distract the wildland agencies from actively managing fire on the land. That September, thanks to the National Association of State Foresters and despite rumored grumbling by his own agency, he was awarded a Golden Smokey Award for fire prevention.

At some point what ignites burns out unless it can find fresh fuel to rekindle. Jack Cohen had always believed that research that failed to connect lab with field was unusable; now he found that research that contradicted the field might be unwelcome. The only solution, he felt, was to return to fundamentals.

———

The nonessentials fell to the wayside. He was asked to leave the National Wildfire Coordinating Group working party; pulled back from counseling Firewise, for which the National Fire Protection Association had assumed responsibility; and after the Angora and Grass Valley fires, quit touring WUI sites, from which he concluded he had little more to learn. Instead he joined colleagues like Mark Finney in a quest to unravel the mechanics of crown fire behavior, which most fire officers regarded as the big fire problem of wildlands, the one that threatened fire crews, inflicted the gravest damages and ran up costs, and, rightly or wrongly, were perceived to power megafires. Once again, he was starting over.

The research questions were the old ones: how did such fires start and how did they spread? But there were wrinkles involving live canopy fuels, which did not burn like saturated dry fuels but more like popcorn kernels (one reason crown fires roared). And there were surprises regarding the difficulty of getting pine needles to burn at all through simple radiant flux. Like fins on a radiator, needles accepted heat, and also dispersed it. What mattered was direct flame contact. But then Jack Cohen knew that. It had been the story of his life.

Ignition had informed his scientific biography and whatever else of his life underwrote that narrative. He knew that big burns could start from a single ember, that great research programs could kindle from a solitary spark of insight, and that a reformation of practice could spread from the glow of an indomitable personality. Of course he understood well that the real power of fire was its power to propagate, and that a start could only spread if it had a suitable setting. But what he learned over the course of his career was that propagation was itself a process of rekindling, and that a lot of embers—or a really determined one—will always find a way.

FIGURE 1. Smoke plume rising from the Selway-Bitterroot Wilderness, 2013. A natural fire left to run its natural destiny. Photo courtesy U.S. Forest Service.

FIGURE 2. Nicholson adit in July 2002. Today an interpretive trail leads from outside Wallace, Idaho, and the entry is cleared and blocked with bars.

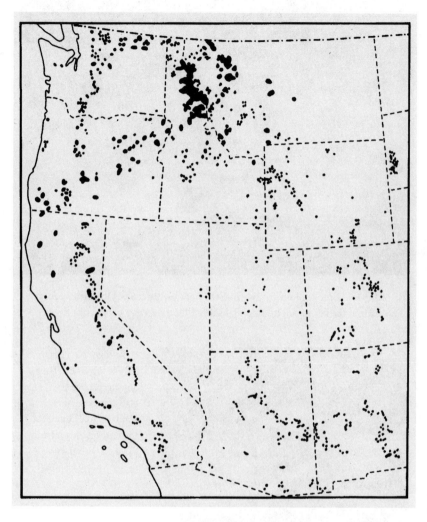

FIGURE 3. Map of western U.S. fires during 1910. The large blob in the
Northern Rockies marks the Big Blowup. Source: U.S. Forest Service.

FIGURE 4. Glacier National Park, east entrance, burned. All Glacier's
entries are similarly burned, save for the inholding at West Glacier.

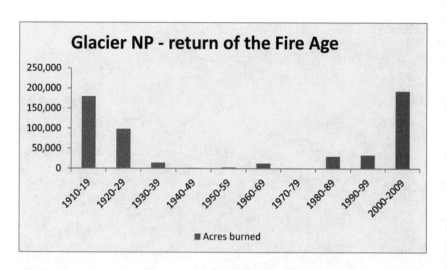

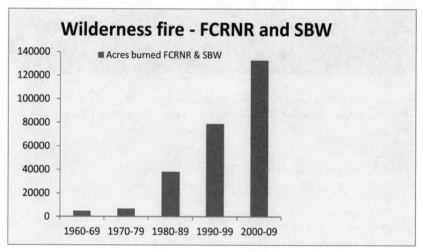

FIGURE 5. Glacier National Park and Selway-Bitterroot Wilderness: burned area by decade. The curve is typical of Northern Rockies landscapes. Fire control was initially effective, but by 1970 an inflection point was reached in which climate and fuels tipped into a future of resurgent big burns. The SBW encouraged the process; Glacier National Park found a way to live with it.

FIGURE 6. Mann Gulch, from the northern ridge, looking southwest toward Helena.

FIGURE 7. The Bad Luck fire: the first official natural fire by the U.S. Forest Service. Photo courtesy Bob Mutch.

FIGURE 8. Centennial commemoration: the rededication of the Big Blowup's firefighter deaths at the St. Maries, Idaho, Woodlawn Cemetery. U.S. Forest Service honor guard, with Fire and Aviation Management Director Tom Harbour (hatless).

FIGURE 9. Bob Mutch (left) and Dave Campbell at the 40th reunion of the Bad Luck fire.

FIGURE 10. Jack Cohen. Photo courtesy U.S. Forest Service.

PORTAL

GATES OF THE MOUNTAINS

ON JULY 19, 1805 the Lewis and Clark expedition, moving up the Missouri River from Great Falls, met the Rocky Mountains. The river breached the high cliffs with, in William Clark's words, barely enough room to place the sole of a moccasin between the water and the stone. The passage alternately widened and shrank for almost six miles before squeezing through a final cleft. Meriwether Lewis named it the Gates of the Rocky Mountains. When they cleared it, they had entered the Rockies proper.

More than anything else, Lewis and Clark defined how the United States would relate to the Northern Rockies. There was the river, the mountains, the game and the furs, and later, the exposed rock that would underwrite a mining economy. But because of hostile tribes the Missouri proved troubling for successors like Ashley's Rocky Mountain Fur Company, and they found alternate routes overland. South Pass—its low dome the geographic antithesis of a gorge—carried that traffic as it later did the overland trails and the transcontinental railway. By Long's 1819 expedition steam was replacing muscle on the river. By midcentury steam overland powered the region's commerce and defined its links to the national narrative. By the onset of the 20th century the defining features shifted to a protected public domain, mostly ensconced in forest reserves and parks.

So while rock and water in its various forms continued to shape the geologic contours of the Rockies, the nation's engagement shifted to forests and fire in its various manifestations. Lewis and Clark said nothing about burning as they sloshed through the Gates of the Mountains—they could not in any event see over the towering cliffs—but free-burning fire had sculpted the looming ecology as completely as the Missouri River had the gorge. The shared history between water and fire is striking. The Progressive Era sought to constrain both in their free-flowing forms. The river was dammed, its torrents tamed into lakes, its energy funneled through turbines. Hauser Dam, south of the Gates, was erected in 1907, failed the next year, and was rebuilt in 1910 (the year of the Big Blowup). Holter Dam, north of the Gates, flooded the Missouri through the mountains in 1918 (a year before the next monster fire year).

Those same sentiments, directed toward fire, favored suppression, followed by a breakdown during the Big Blowup, followed by a vigorous reconstruction. More deeply they pointed toward the abolition of wildfire in favor of internal combustion, which promised to do for fire what hydrodynamos did for water. As the century evolved industrial burning began a curious dialectic with wildland flame, and the use of concentrated firepower in machines was the primary counterforce to meet free-burning combustion. For a while it seemed that the region's fires, too, might be dammed, no more given to blowups than its rivers to floods. But water is a substance, fire a reaction; rain can be impounded, but fuels only stockpile; eventually they burst their barriers. Reservoir sites can hold the runoff of watersheds for useful storage and later use. Forest reserves, however, if denied regular burning, only stoke conflagrations.

Then the wild became wilderness. The Wilderness Act redefined not only what people might do on land set aside under its provisions but how they understood those places, and what narrative they might apply to them. The Gates of the Mountains remained a marvelous entry point not only because most of it was inaccessible in any practical sense to logging, mining, grazing, or roaded traffic, and hence easily reclassified as wilderness, but because it could update the prevailing narrative of what a wilderness encounter ought to mean. Wilderness was a story as well as a site: it spoke to an American epic, a national creation story, in which the civilized met the wild and found itself renewed. The Gates were a portal to that journey. There were places more remote, more spectacular in scale, more panoramic with scenic awe. But there were few if any that combined such a setting with a narrative of equal power. Unsurprisingly, the Gates of the Mountains were among the first places identified for inclusion under the Wilderness Act.

As the wild became valued, so the understanding of its waters and fires shifted as well. Free-flowing rivers became places to preserve rather than impound. Free-ranging fires were events to accommodate instead of suppress. Formal designation as wilderness gave legal leverage to allow that shift. What the new era needed, however, was not just a suitable site but a usable story. What keeps the Gates at the center of this new era is that it supplied that narrative.

It happened at an otherwise inconspicuous ravine that spilled into the Missouri's gorge just past the Gates proper. The story of the Mann Gulch fire of 1949, once it was written by Norman Maclean in 1992, gave the country its most compelling fire narrative since the Big Blowup. It provided a means by which the nation might reenter the region—might understand the character of the place and the country; might appreciate what it had to say about America's encounter with its lands; might understand fires that otherwise belonged with beetle outbreaks and blizzards.

In 2007 the Gates of the Mountains Wilderness burned over. Its waters might no longer flow through the mountains, but its fires rushed unimpeded over them.

YOUNG MEN, OLD MEN, AND FIRE

You picture the mountainside as sides of an amphitheater crowded with admirers, among whom always is your father, who fought fires in his time.

—NORMAN MACLEAN, *YOUNG MEN AND FIRE*

FROM BILL BELL TO BUD MOORE

[Bud Moore] and I soon discovered that both of us had worked in the Lochsa when we were boys and when the Lochsa was thought to be accessible only to the best men in the woods.

—NORMAN MACLEAN, *YOUNG MEN AND FIRE*

N 1930, AS BERTRAND RUSSELL readied for publication his specu- lations about a human future driven by applied science, he inserted a counter observation from the long past. At issue was the concept of "state of nature." The view that "man should live according to nature" was a conception that is "continually recurring throughout the ages," he intoned, "though always with a different connotation." In practice, he concluded, "return to nature" meant the conditions that the writer knew in his youth. That same year a 12-year-old Bud Moore, on a solitary trek, crested the Bitterroot Mountains and looked across the landscape of the Lochsa Valley. It was a panorama that became a vision that evolved into an epiphany. The Lochsa would inform his life.[1]

Bud Moore saw his first fire when he was six. His father extinguished a nearby snag kindled by lightning and was paid by the Forest Service. Young Bud fought his first fire four years after his defining Lochsa trek. He imagined rangers like the celebrated Bill Bell as the logical successors to the free-spirited mountain men he admired as a youth. It was a way to live within the wild. The Forest Service had a founding legend, a camaraderie, and a code based on toughness with a call to duty that made its rangers the offspring of the western hero. All this showed itself most spectacularly in their astonishing fight against wildfire.

In 1928 he was hired to work on the Powell District, mostly trail and telephone line maintenance, and of course on fires whenever they popped up. Smokechasing was a mainstay of life on the Lochsa. He met many of the legends of the Northern Rockies—Bill Bell, of course, but also Ed MacKay of the Powell, Elers Koch, and Major Evan Kelly. They were backcountry and fire men all. He was on the line during the big Selway fires of 1934. When World War II broke out, MacKay and Kelly recruited him to help with the guayule project in Southern California. While there Moore enlisted with the Marines, where he found himself again fighting fire at Camp Pendleton before heading to the Pacific. When the war ended he returned to the Forest Service with a war-service appointment and was assigned as an alternate ranger to the Powell, where he mostly fought fire.

Bud Moore was an American type, the self-educated boy from the frontier whose grit, talent, and instincts allowed him to rise through the ranks. His stroke of fortune came when, soon after his return, he was "grandfathered" into a "professional" appointment as an assistant ranger without the expected education. He oversaw the postwar development of the northern Lochsa. Then he joined the committee of inquiry that looked at the string of tragedy fires in Southern California, in which he returned to his Marine Corps training and restated the Corps's standing orders into the 10 Standard Fire Fighting Orders. He went to the Washington Office as an assistant training officer, became more fully involved with fire (which is where most of the training belonged), and joined the National Fire Coordination Study that surveyed America's fire panorama for the Office of Civil Defense. In 1967, as big fires swept the Northern Rockies, Moore reviewed the scene for the national office and reaffirmed the value of initial attack. Soon afterward he became deputy national director of Fire Control, and then director. Finally, yearning for the homeland of his youth, he returned to Missoula as fire director for Region One. When he retired in 1974 he probably knew as much about fire across the country, and from the ground up, as anyone in the USFS.

In his final tour Moore appreciated that the old ways couldn't continue. He distrusted the analogy of firefighting to battle; he knew the

difference, disliked the Marine Corps's singularity of focus and reduc-
tionism of everything to the single task at hand, which might work on a
battleground but could only fail in complex landscapes. He recognized
that fire couldn't be stopped and that in many settings more fire not less
was needed. What tipped the scale was the Wilderness Act. He harked
back to his youth and that epiphany on the Bitterroots and decided that
"nothing is more needed in wilderness than fire," certainly in the Northern Rockies.

Perhaps, too, he saw the restoration of a natural process as partial
redemption for what he had done as ranger on the Powell District, for he
had helped introduce the toxic worms that had eaten into the wild apple
of the Lochsa. When a spruce budworm epidemic broke out, the proposed remedy was wholesale logging, so he watched bulldozers achieve
for timber companies what the Nez Perce, army engineers, and railroad
magnates had failed to do: open the Lochsa Valley to active exploitation.
It was one thing to hunt lynx and marten along traplines accessible only
to snowshoes. It was something else to push unstable slopes into once-
clear trout streams and fell whole hillsides. When the axe failed to keep
up with the insects, the agency turned to DDT. Throughout, the Forest
Service continued, at enormous labor, to fight fires. In the postwar era
aircraft became an indispensable part of its armory as fire officers sought
to reach ever more remote fires sooner and with ever greater power. The
aerial firefight was the mechanical equivalent to those dozer roads crash-
ing through the wilderness.

Bud Moore had begun to doubt. Perhaps fires were no different from
free-ranging grizzlies or wolves, and the countryside was the worse for
their absence. Besides, after 60 years of attempted suppression, and 35
years since the 10 a.m. policy had been promulgated in large measure to
control the big burns of the Rockies, the policy had failed. It was expensive. It was dangerous. It was self-defeating. The fires would come. The
more they were held off, the worse their sweep when they eventually kindled. He had spent his entire life fighting fires, and still they came. It
seemed the agency might be destroying what it sought to save. Maybe fire
was not an enemy to be annihilated but a storm of nature to be accommodated and weathered like blizzards and droughts. When the wilderness
movement arrived, Moore felt a kinship with its ideals. The upshot was
the White Cap Project, chartered in 1970 for the Selway-Bitterroot Wilderness; it received its first natural fire in 1972. Two years later the Forest

Service officially retitled its Division of Fire Control as the Division of Fire Management. With that it was time, Bud Moore decided, to retire.

In New England, when you wish to proclaim your status and retire in the style of the countryside, you buy a colonial farmhouse. In the South, you buy a plantation. In Texas, a ranch. In the Rockies, you get a cabin. Bud Moore did one better. He built his out of logs by hand, returning to the world he had known as a youth. He then turned his vision into a philosophy of land ethics. He sought to replace the timber cruising that had ripped open the Bitterroots with "eco-cruising." He wrote a book. He became for a new generation what Bill Bell had been in his own youth, the beau ideal of the ranger. He was the man the next cohort looked to for insight and approbation. And while the University of Montana awarded him an honorary doctorate, a triumph for a man whose schooling had ended in the eighth grade, probably the greater pleasure came when the White Cap Five convened around a campfire on Cooper's Flat in early September 2002 for the 30th anniversary of fire's reintroduction to the greater Lochsa.

The restoration had been, in the deepest sense, an ethical act, and it had been one designed to pass the torch from one generation to another. It was not only about the wild but about initiating the young into it. "To me," Bud Moore confirmed at his retirement, "most of all the Forest Service is the eager uncertainty of young men and women as they confront an old pro at their first job in the woods."[2]

FROM BUD MOORE TO BOB MUTCH

Could you expect less from a boy who grew up in the woods and grew old as a schoolteacher and so spent most of his life staying close to the young who are elite and select and, by definition, often in trouble? I came to Mann Gulch expecting to catch glimpses of them as far as they could go. That's why I came.

—NORMAN MACLEAN, *YOUNG MEN AND FIRE*

It's a long way, in more than geography, from Cleveland, Ohio, to Paradise Guard Station in the Selway country. But it's a leap Bob Mutch made first in his mind as a boy, then as a smokejumper, and finally as lead scientist on the White Cap Project. What imprinted on his childhood memory were the woods around that gritty city. What initiated him into his mature years were wildfires in the Northern Rockies. They merged, as they had for Bud Moore, into an epiphany that became an ethos.[3]

From Cleveland suburbs Bob went to Albion College and decided he wanted to be a forester. In 1953, at the age of 19, he joined a blister rust crew at St. Maries, Idaho, close to ground zero of the Great Fires of 1910, and it was fire that "rescued" him from a grueling summer of plucking *Ribes* like gooseberry along endless transects. That first fire demanded a long hike into the Salmon River country and concluded with a magnificent panorama. He tasted his first coffee, served by logging crews impressed for fireline duty. He met some smokejumpers from Missoula. The next year he joined them—a member of the first class in the new Aerial Fire Depot, which President Eisenhower dedicated that September. His two years on the cadre were slow, but Bob made his first jump, the Ballinger Point fire, in the Selway-Bitterroot primitive area. He was hooked. He enrolled at the University of Montana for a graduate degree in forestry.

He moved into research. He worked at the Priest River Experiment Station—Harry Gisborne's stomping grounds—and when the Missoula lab opened in 1960, he was among its first hires. Meanwhile, he remained active in making knowledge relevant to fire protection by serving as a fire behavior officer on an overhead crew. But the sense gnawed at him that something was missing, that the era of fighting every fire everywhere and of ramping up research to help fight them and even (among one of the founding objectives of the Missoula lab) trying to suppress lightning in order to stop ignitions, could not continue. Something was fundamentally out of whack.

His personal annus mirabilis came in 1970. He published in *Ecology* an article—his most famous—that reversed the usual conception of fire adaptation. The Mutch hypothesis argued that plants not only adapt to protect themselves from fire, but some can be seen to promote fire; that flammability, paradoxically, can confer selective advantage to those plants better adapted to recover from them. The subtext was that fire is not just something out in nature like ice storms or floods against which species must shield themselves, but something that has emerged out of a long coevolution to which plants themselves have contributed. Fire was not simply the outcome of climate and fuels but expressed complex biological processes. The upshot was, suppressing fire is not only unnatural but disruptive to precisely those species that have most accommodated it. At the same time, Bob saw a flyer at the lab soliciting a "wilderness planner" for an experiment in natural fire management. His application went to Bud Moore, by that time Region One director of Fire Control. Bob Mutch joined Dave Aldrich to create the plans behind the White Cap Project that would begin restoring fire to biotas that needed it. If insights were to have meaning, they had to find expression on the ground.

The two men traveled to Sequoia-Kings Canyon National Park to see firsthand the earliest trials of letting fires burn. In 1971 they joined Moore, Bill Worf, and Orville Daniels—what became known later as the White Cap Five—on a trek into the Selway-Bitterroot Wilderness to discuss their conceptions on site. The following summer the program got approval as an exception to the 10 a.m. policy; the next day the Bad Luck fire provided nature's imprimatur; a year later the Fitz Creek fire blew up, tested ideas and resolve, and confirmed the program. The year following, having completed what he regarded as his life's task, Bud Moore retired. In 1977 when fire research sought to relocate him to the Riverside lab, Bob balked, and instead transferred to the fire staff of the Lolo National Forest. When, the next year, the Forest Service officially renounced the 10 a.m. policy, he was in a position to advise that fire-rich forest with fresh plans. He was where he wanted to be, the place that most melded head, heart, and hands.

When he left, he traded the Bitterroots for the world. In 1986 he accepted an assignment in the Washington Office as the first program manager for the Disaster Assistance Support Program, through which the Forest Service satisfied requests from the State Department. He became, in effect, America's incident commander for responding to fire and other emergency requests throughout the world; but it was, as Moore's first Washington Office assignment had been, an educational as much as an operational mission. For five years he set up train-the-trainer programs, oversaw specialists to help control locusts, counter famines, respond to hurricanes, and of course cope with wildfires; fire specialists or equipment went to Latin America, China, and even the Galapagos. In 1988 he self-deployed to the Yellowstone bust, effectively bringing the world home. Throughout, he was once again translating knowledge into practice. He continued for five years, then recycled back to the Missoula lab as a specialist in technology transfer. From time to time that included overseas assignments.

It was in Brazil in 1994 that it all came together. Suddenly, while drinking beer with a Brazilian *bombeiro*, he had another epiphany that he believed "could also appear as [his] epitaph." His life's purpose could be summed up in six words: "finding harmony among people and ecosystems." That meant putting fire back where it belonged, stopping fire where it didn't, and keeping those who managed fire safe. The discovery brought "a sense of closure and satisfaction" to his career. Although the terms had morphed—"land" had become "ecosystems," and "technology transfer" had replaced "passing along what you know"—the sentiment was nonetheless interchangeable with Bud Moore's vision, or with Bertrand Russell's. It conveyed both a mission and a morality. That year Bob Mutch retired.

He remained active, opening up a consultancy, accepting assignments from the World Bank and the UN Food and Agriculture Organization that took him to Bulgaria, Ethiopia, India, Mongolia, and (many times) to Brazil. He spoke often at training sessions and conferences. But his most fulfilling moment, hands down, across his 59-year career in fire was his work in the White Cap where he harmonized, at least in principle, people and ecosystems. Before retiring he bought land outside in the West Fork and built a cabin. In 2002 he joined the White Cap Five for a 30th anniversary of the Bad Luck fire. A decade later he returned for the 40th. By then, the torch had passed beyond the West Fork.

BEYOND 40

It is a great privilege to possess the friendship of a young man who is as good or better than you at what you intended to be when you were his age.... It is as if old age fortuitously had enriched your life by letting you live two lives, the life you finally chose to live and a working copy of the one you started out to live.

—MACLEAN, *YOUNG MEN AND FIRE*

That a man might treasure the moment of his initiation into manhood is common enough, especially if it involves a hard physical trial, and there is little novel if he should identify it with a place, particularly one as overpowering as the Selway-Bitterroots that can boost a routine coming-of-age saga into a vision quest. That it might crystallize into a code for living is less typical, but far from rare. What Bud Moore and Bob Mutch and uncounted others experienced has counterparts elsewhere throughout the fire community and beyond.

What distinguishes the Northern Rockies is not the individual moment of revelation but its transfer across generations. The Southeast with its tradition of handing down burning from parent to child has a comparable social character. Many California fire officers recall fondly how their fathers took them out after Sunday supper to look for elusive smokes and put a pine bough in their hands to help swat it out. Still, its intergenerational theme (even across cultures) may be the most unusual feature of the Lochsa story and of fire in the Northern Rockies generally. It reverberates in campfire stories, memoirs, and ceremony; and it inscribes an interlinear text in the region's most prominent fire writer, Norman Maclean. "USFS, 1919," one of the tales in *A River Runs Through It*, is his coming-of-age story under the gaze of Bill Bell. *Young Men and Fire*, his meditation on the 1949 Mann Gulch tragedy, is about old men as well as young, and the connections between them as the young man ages, and the aged author tries to explain something about the universe. The book opens with Maclean's youth, then seeks to rescue the memory of the youths lost at Mann Gulch, and throughout accents his relationship with Laird Robinson as they pull each other along, one relying on the wisdom of more than 70 years and the other on the residual vitality of youth. The cycle of fire in the Northern Rockies involves generations of people as much as scorched conifers.

Often the handover occurs within families. The federal land agencies have long displayed a quasi-caste quality akin to military families in which children, having grown up on bases, follow their parents' career. The record of second- or third-generation fire officers is striking. It may be that the region's generational theme will become still more genealogical, particularly after the Forest Service underwent a wrenching court-mandated demographic shift as a result of the 1981 consent agreement that brought in large numbers of minorities, notably women. The agency had to find new ways to move people rapidly through the ranks; the newcomers were often older, better educated, and not drawn from traditional or common pools of experience; they could not wait for the inherited processes of initiation. The saying in Region One used to be until you'd been on a hundred fires, you kept your mouth shut. That would not be possible with the volume of workforce turnover. Instead, continuity may come through family lines rather than bureaucratic ones.

Maybe, or maybe not. For a hundred years, however, the lore of the Lochsa has passed from old to young. In 1971 Bob Mutch began taking his children—Dale, Brian, and Linda—with him when he backpacked into the SBW. His daughter, Linda, was 10 when she first trekked with him. Growing up she studied fire, spent summers on fire crews when she went to college, worked two seasons on fire-related projects for the Park Service in Alaska and then in Sequoia-Kings Canyon's Big Trees before migrating into wilderness inventory and monitoring. At the 30th anniversary of the Bad Luck fire, she spoke about how the trails taken in the White Cap had become a career path. And when in the summer of 2012 the Salamander fire free-burned on the Selway-Bitterroot Wilderness, Bud Moore's son, Bill, was a volunteer staffing the Salmon Mountain Lookout to report its movements.

HOW I CAME TO MANN GULCH

In my story of the Mann Gulch fire, how I first came to Mann Gulch is part of the story.

—NORMAN MACLEAN, *YOUNG MEN AND FIRE*

WHEN NORMAN MACLEAN was writing *Young Men and Fire* he says he was on the downhill side of his Biblically allotted three-score-and-ten years. When I came I was on the uphill side, with a few years yet to go, though I was aged enough to allow for a pilgrim's staff. I had mistakenly forgotten my maps, but didn't really need them because I was not there to recreate the fire and because the landing at Meriwether Canyon was closed due to recent postfire debris. I was dropped directly at Mann Gulch. There was only one way to go.

The place has a preternatural quiet that shimmers between the sinister and the hallowed. The hum of flies and bees, the chirp of an occasional bird, the hush of the wind; there is nothing more. Completely absent is the secondhand babble that engulfs modern life. It's been said that exploration without an intellectual purpose is just adventuring, that fasting without prayer is just going hungry. So solitude without contemplation is just lonesomeness. Mann Gulch is a place that invites thought.

It is a pilgrimage site, one of the few the wildland fire community has. There are memorials and crosses where the 13 smokejumpers died and a path that remains visible though not formally maintained. Visitors leave small piles of stones, rather like Buddhist pilgrims tying a piece of cloth.

Mann Gulch is remote and obscure enough to filter out the casual tourist and curiosity seeker. It is common for fire to be associated with some spiritual interest. The oddity at Mann Gulch is that the spiritual interest is the fire itself.

You hear often enough from the fire guild about the lessons to be learned at Mann Gulch, as with all tragedy fires. You will sometimes detect sneers from hardcore firefighters that the 1949 jumpers made mistakes that proper training would never allow today. You may be lectured by fire behavior analysts and fire scientists about what caused the blowup that consumed the side-canyon and how those on site might have evaded their tragic fate. But trekking into Mann Gulch to discuss fire behavior is like taking the pilgrim's trail to Santiago de Compostela to discuss catechisms. It's an error of literary, and even moral, judgment far more egregious than anything the jumper crew putatively did on August 5, 1949.

My reason for coming was not to honor those who died, although I would do that, nor to ferret out or puzzle through missing mechanisms of the fire, although the 360-degree panorama of the site made crystal clear what had happened and why. Rather, I came to pay homage to Norman Maclean. I wanted to know how you create a literary blowup.

━━━━━━━━━

Where now is our Tolstoy, I said, to bring the truth of all this home. . . .
Must we wait for some one born and bred and living as a laborer him-
self, but who, by grace of Heaven, shall also find a literary voice?
—WILLIAM JAMES, "WHAT MAKES A LIFE SIGNIFICANT?"

Big fires require many coincidences, or what Maclean in his crusty woodsman's persona called screwups. The day was hot, the air unstable, the fuels ripened over a long summer. The winds of the Missouri gorge met those of the gulch and the plume of the building fire. So, too, tragedy required other screwups that put the wrong number (or right) in the wrong place (or right) at just the wrong (right) time. The crew had to arrive neither too late, nor too early, which meant they had to parachute in, and then they had to find themselves too high up to get below the fire and too far down to outrace it up the slopes. Fire and fire crew had to converge with only one possible outcome.

Yet the same holds for great works of the imagination, those that are not merely clever but that exhibit the fundamentals of the human spirit as big fires do the physics of heat transfer. They, too, involve a convergence of screwups. The place has to fit the story, the story has to fit the theme, the text has to reconcile style with subject, and then the book has to bond with the larger culture. It was Maclean's genius to find the perfect place and story, and then to blend them in an evocation that only the most obtuse, literal, and positivistic could find objectionable. He understood that big fires could be tragedies, and that tragedies behaved, as a literary creation, much like big fires. They blow up. It was his triumph to have that work subsequently speak to the American fire community and then to American society in ways no other writing has.

How this happened is neither obvious nor inevitable. Mann Gulch had been in the chronicles for 28 years before Maclean took up the cause. Prior to 1992, when *Young Men and Fire* was published, Americans had had a century of forest reserves, on which they fought fires, without creating more than the odd doggerel and Sunday supplement journalism. Until Maclean saw the "It" through the smoke hardly anyone appreciated the literary potential of fires in a Montana mountain. But it is equally true that it took Maclean, who claims to have visited the area while it was still smoking, a long time to appreciate the possibilities. For decades he tried to write what would become the informing themes of *Young Men and Fire* by using the massacre at the Little Bighorn. He couldn't make it work. At Mann Gulch everything—all the little screwups of setting, symbol, and story—converged.

Consider each part in turn. First, the place. Mann Gulch presents a self-contained world, though one inverted from the usual conception of mountain ravines. It is wide at top and narrow at bottom. During the fire no one could get into it from below (the one man who tried passed out); and only two got through openings at the top (for reasons they could not explain). The sweep of rounding ridges along the crest top form a broad amphitheater, and Maclean sketches his jumpers as insouciantly imagining themselves performing for a crowd. Geographically, the Gulch is a perfect microcosm. It contains all the action.

Next, the story. It, too, is a miniature of a life cycle. The jumpers are born out of the sky, leaving the womb of the C-47 and breaking the umbilical cord of their static line as they descend. They live briefly on the slopes as they move to what they assume is a normal destiny. Then the fire blows up below them. They die, save for three. There is a remarkable unity of place and action, as Aristotle might have put it. The story is as self-contained as the place.

Still, it would have been unusable, as the Custer massacre was, without survivors. The smokejumper corps believe themselves an elite, which Maclean recodes as Calvinist elect. They are set apart from the run-of-the-mill populace who shun fires or find themselves recruited off bar-stools. Two smokejumpers survive for reasons that no one can discern other than chance or predestination. The Mann Gulch blowup, as scientists from the Missoula lab prove with mathematical rigor, was a "race that couldn't be won." The natural order determined to clean out the gulch. Firefighters, even those who thought themselves spared from the mundane and the mortal, had no more say than squirrels and deer.

The critical action turns on Wag Dodge and his "escape fire." And an escape it truly is. It allows him to evade the foreordained fates for all those seemingly destined for the flames of perdition. Through his own hands and wits he saves himself, and it is Maclean's challenge to show that in saving himself he did not doom others. Dodge is the hinge on which the drama turns. A story of action, however graphic, cannot carry much moral burden. A story without choice or agency is natural history, not literature. Without Dodge's escape fire, the saga of Mann Gulch is an accident. With Dodge, it is tragedy.

Still, the text did not write itself. Tens of thousands of people have been on firelines, and thousands knew about Mann Gulch. The after-action studies led to reports and anecdotes and a Board of Review. But nothing like literature happened until Maclean did it. He moved from campfire anecdote to parable and then to universal tragedy. He saw in the episode what others did not. He wrote the equivalent of an escape fire.

The rules for nonfiction are simple enough. You can't make stuff up, you can't leave out something that truly matters, you have to match style and subject. Maclean's purpose is story, but narrative of a peculiar sort, because, in the end, the story is about storytelling. There is no formal outcome, tested as true and confirmed, a data set akin to a laboratory experiment. In fact, Maclean tries one form of explanation after another and they all fail except for simple storytelling, of the search as itself a story. So the design wanders, or appears to (Maclean is too savvy for a genuine random walk). What will make or break the tale is the voice of the storyteller.

Shrewdly, Maclean adopts the persona of an old woodsman, a man who knows fire and the countryside and the people who work it. In truth, he hadn't been on a fire since, at age 17, he had labored on a Forest Service trail crew. So he's not exactly a Tolstoy, narrating omnisciently, but neither is he a laborer who got educated into literary theory. The fancy concepts get buried, as the old woodsman allows Maclean to indulge in plot diversions and slang and a colloquial perspective that keeps the storytelling front and center. The upshot is not the true story of Mann Gulch but a true quest for the storytelling, which is the best someone can hope for. So place and event merge, like a reverse prism in which multiple bands refract into a single white light.

That leaves a bond with the culture because a book does not contain its own meaning but like a fire is a reaction that derives its significance from its setting and whose power lies in the power to propagate. For a blowup book there must be some triggering fire whirl of meaning. Maclean had a leg up because he was already a celebrity writer with a national audience thanks to *A River Runs Through It* and because his pursuit of the Mann Gulch story was widely known among the cognoscenti. In a perverse way his death before completing the text added piquancy. Like his portrayal of the jumpers who lurched ahead after the flames had overtaken them, *Young Men and Fire* seemed to carry Maclean's last wishes a final lunge after death had overtaken him. The book became a bestseller and won a National Book Critics Circle award.

Public attention, however, was framed by two events. In 1988 Yellowstone had burned (and burned and burned) and alerted the American public that there might be more to fire than a day or two's flash of flame, and prepared it to move from fires etched into a celebrity landscape to a fire from a celebrity author. The defining event came two years after publication. The South Canyon fire in Colorado seemed, with unsettling fidelity, to reenact the tragedy at Mann Gulch. For decades fires had burned over crews, and if not indifferent, the culture had little mechanism to interpret those losses beyond a regrettable industrial accident.

Other than a spread in *Life* magazine, little came from Mann Gulch. It shot up a photogenic plume, which then blew away with the next day's news. Thanks to *Young Men and Fire*, however, the public had in its hands a Rosetta stone by which to read the hieroglyphics on the slopes of its successor at Storm King Mountain.

The fire, part of a ruinous season that killed 34 firefighters and burned through nearly a billion dollars in suppression funds, galvanized the fire community. Within a year a new common federal wildland fire policy was announced along with protocols to help ensure the safety of crews. The book and its times bonded, or as Maclean might have put it, the little screwups of the inside world that was the book began to fit with the screwups of the outside world that was American society until they became the same. Without *Young Men and Fire*, the reforms would likely have come piecemeal and more haltingly. And without a recapitulating fire to confirm it, *Young Men and Fire* would have passed, not unlike the fire of August 5, 1949, without much purchase. The downside has been the propagation of cloying imitators, like a field of thistle, and the preciousness that has attached to everyone who dons Nomex. None of that can be charged to Maclean.

What he can claim is a singular success at changing how Americans thought about wildland fire and firefighters. The fact is, the Mann Gulch fire barely registered beyond the Northern Rockies. It was even possible, until 1992, to write the fire history of the country with hardly a mention beyond including it in the roll call of tragedy fires from 1937 to 1956. The tally of what it did not do is longer than what most partisans believe. It did not redirect fire research. A reorganization was already underway dating from 1948, and what galvanized research into fire behavior was not a blowup fire in a small side-canyon of the Rockies but the explosion two weeks later of an atom bomb by the Soviet Union. (Almost contemptuously the fire claimed the life of Harry Gisborne, the leading regional fire researcher, who collapsed while tramping around the site to investigate what had caused the blowup.) It did not refashion training and fireline operations. That came after two more, larger tragedies in California. It did not implant fire as a plausible theme within the larger culture despite a photo-essay in *Life* magazine and a Hollywood movie, *Red Skies of Montana*, obviously based on it.

For American society wildland fires remained a curiosity and freak of western violence, like a grizzly bear attack. The impact of Mann Gulch remained stubbornly regional, agency-specific, even personal; the smokejumper corps, Forest Service Region One, and a handful of local oracles kept the memory alive. It inspired no memorial services, it sparked no literary legacy, it came and went like the fire that drove it. There are hints that the Forest Service, at a national level, was content to have it forgotten. Nor did Mann Gulch focus a national reconsideration of how the country ought to engage wildland fire. In 1949 the United States had endured a horrific world war, had entered a cold war, and, though unaware, was poised to enter another hot war in Korea; and since firefighting was war by other means, casualties would happen. No one wanted them, no one planned for them, they were just part of the business, or what the pretentious called "battling the red menace."

The fact is, the Mann Gulch fire was trapped in an obscure side-canyon of history until Norman Maclean jumped on it. It was his literary imagination that stirred and scattered the embers into a blowup. Almost single-handedly *Young Men and Fire* established a literary genre, shouldered its way into the consciousness of the fire community, and more astonishingly bestowed onto American fire what it had never had before, a kind of Tolstoyan saga that gripped the public at large. It gave the literate public a kind of reading glasses with which it could see the 1994 South Canyon fire as something other than, yet another blurry entry in, a long fuzzy chronicle to the status of cultural catalyst. No other piece of literature in American fire history has come close.

In a story compounded of paradoxes, this may be the most telling: over the course of the fire revolution the most significant publication— one of the revolution's decisive moments—did not come from forestry schools or fire labs but from a professor of Renaissance literature at the University of Chicago. Other fire intellectuals knew how to put numbers on the phenomenology of fire. Maclean knew how to create meaning out of it.

Ultimately the Mann Gulch fire did all that its recent chroniclers have claimed for it. Those effects just happened not in 1949 but in 1994, two years after Maclean explained why we might care.

But why did I come? The Mann Gulch fire occurred almost exactly five months after I was born, and *Young Men and Fire* had been published twenty years in the past by the time I trekked up the slopes on July 9, 2012. A decade earlier, in *Smokechasing*, I had published a critical essay on the book as literature, and had little to add. What brought me was the sense that American fire needed a new narrative to explain what was happening on its lands and to its culture. I wanted to know how a narrative became great. I had earlier written about the Big Blowup with that quest partly in mind, and now I wanted to understand better what alchemy had made *Young Men and Fire* possible.

The issue was not academic. In 2007 the Meriwether fire burned over the entire Gates of the Mountains Wilderness, including Mann Gulch. The river passage to the mouth of Mann Gulch was lined, on its eastern flanks, with burned hillsides and scorched trunks and trees with blackened catfaces along the river's edge. The new fire had reburned not only the scene of the 1949 tragedy but the entire range. In a real way it had also burned over the story of *Young Men and Fire*; so while scars from the 1949 fire yet remain, the scene has been erased and written over like the ink in a palimpsest. The book endures as art, but the template it added to the narrative cache available to the American fire community may not apply 15 years later.

The Meriwether fire began when lightning kindled a fire on the ridge between Colter and Meriwether Canyons on July 17, 2007, as part of a regional bust. There were more compelling starts than one in a wilderness, and the specter of Mann Gulch still haunted the place and the tragedy's host forest was not about to send crews in to fight it. For several days they watched and occasionally drenched the creeping burn with helibuckets while they threw crews and engines at fires scattered like errant comets around Helena. Somehow—it's possible that one of those water dumps or the rotor wash from a Skycrane blew sparks down the mouth of Colter Canyon or roused the slumbering flames to fury—the fire blew up on July 23, a mechanical recreation of the whirl that had perhaps eddied from the shearing winds between the Missouri gorge and Mann Gulch. The flames surged over Cap Mountain, then galloped across the entire Gates of the Mountains Wilderness.[1]

This time there were no crews strung out along the slopes and daydreaming or taking photos. Before it ended, the fire blackened 47,000 acres, nearly twice the size of the legal wilderness, and threatened a subdivision along the forest's south boundary, which was spared in part because fuels work had been completed earlier that spring. But no one died. No one trekked over the ridges pulaski in hand. No one subsequently scrutinized the fire's behavior. No one wrote up so much as an official narrative. The fire had passed over the Gates with no more cultural consequence than had it been a flock of mountain sheep.

The Meriwether fire is now the dominant ecological story of fire at the Gates of the Mountains and Mann Gulch. But it doesn't fit any of the existing narrative templates. It was not a disaster story, it was not a classic firefight, and it was not a young-men-and-fire tragedy. It is to our times what the ten o'clock fire was in 1949. If it is to have cultural traction, it will need a narrative that walks with caulk boots up the slopes that matters to everyone outside the fire community.

It won't match *Young Men and Fire* until it again records what William James called "the sight of the struggle going on," of "human behavior *in extremis*," and that will require putting people back into the scene. But the story needn't be tragedy. It just needs to find some way to engage with the larger culture in a morally compelling way and be granted a narrative to carry that meaning. That task will need the right place, the right story, the right imagination, and then the luck of the draw, the right timing to say what the sustaining society wants to hear. The Meriwether fire won't do that, but another fire might, and we can only hope that before too much longer a new Tolstoy will peer through the smoke and find it.

Written on the ridge separating Rescue Gulch from Mann Gulch

WHAT MAKES A FIRE
SIGNIFICANT?

I T ISN'T SIZE. The 1871 fire that incinerated Chicago and provoked panic on the New York Stock Exchange tallied 2,100 acres. The 1906 San Francisco fire burned 3,000. But using urban fires for an index is comparing pulaskis and shovels. Surely, wildland fires have their own dynamic of meaning for which size matters. Well, yes and no.

What makes a fire important is that it connects with the culture in some way that allows the burn to transcend its local setting. Those fires must matter. They must interact with a receptive society. They must get institutionalized, they must be recorded and pondered, they must hybridize with literature, art, politics, the sensemaking of those who live with and around them. Urban fires are positioned where they can produce casualties and be reported, which is to say, they become stories with currency in the larger culture. Wildland fires that become significant are more likely to be big because a big burn is better situated to find some points of contact than a small one, though not all big fires are significant, and not all significant fires are big. A region like the Northern Rockies is likely to claim a higher percentage of transformative fires because it has so many big ones. Even so, vast landscapes can be culturally empty except for those values, like wilderness, associated with emptiness. What matters is where fires and culture meet.

Today, we typically equate "significant" with "celebrity." For a fire to get public buzz it has to burn houses, kill people, or involve celebrities (a celebrity landscape will do). Otherwise, a large fire in a remote, unsung landscape is just a big smoke in the woods. They come and go with no more moment than elk in migration.

But what about in the past? Big fires didn't begin in 1910. The 1825 Miramichi fire across Maine and New Brunswick burned as large a landscape as the famed Big Blowup. There was enough of a jolt that it entered into historical archives and even inspired a folk song or two. The monster fires of 1871, 1881, 1894, 1902, 1903, and 1908 that gorged on the slash left by industrial-strength logging all found their way into the master chronicle because someone wrote about them or consciously created a memorial to them. It helped that whole communities were consumed, that hundreds died, and that these fires might be linked to other historical moments such as the Chicago fire.

What they share is that they all left a story in the ash. They tapped into older plots, contemporary memes, or prevailing metaphors—they illustrated themes and issues of interest to the public, which is to say, an audience. The Northern Rockies has a disproportionate number of such storied fires because it has so many blowups, because the U.S. Forest Service has such a commanding presence, and because the fires wormed into the cultural software of the agency. Once established, the lore of big fires became its own referent.

In the 19th century the significant fires raged through the Lake States. In the early 20th they appeared in the Northern Rockies, and in the postwar era, in California, which came with its own media megaphone. The big-fire era in the Lake States ended when fire and axe gutted its forests; fires would return only insofar as the woods did. In the Northern Rockies the big fires merged in the postwar era into a larger discourse about wilderness. But big fires had happened long before. As soon as the mountains had shed ice for trees, they burned, probably on a massive scale. Where are they?

———

They exist in the historic record, but with few exceptions not much in the cultural one. The obvious geodetic marker for comparison are the fires of 1889. The 1871 Peshtigo fire had joined America's best known fires because its curators have celebrated it as America's Forgotten Fire, which it hardly is. Researchers who have traced the geography of the 1889 fires have attempted to make it also famous for being overlooked. The burning that droughty season certainly ranged far. A historic survey identified some 35 region-spanning "fire episodes" in the Northern Rockies from 1540 to 1940, which left about 12 years between eruptions. There were plenty of other big burns, though they tended to be localized on a mountain range or two, not sprawling over the northern cordillera. Of the whole roster the "circa 1889 fires" yielded "the most compelling evidence of extensive burning."[1]

Mostly, the historic record of that year chronicles the vicious urban fires that ravaged large chunks of St. Louis, Buffalo, Philadelphia, Boston, Chicago, Savannah, Seattle, Spokane, and even New York City. The fires that made page eight were those that rambled in New Jersey, Virginia, Wisconsin, Michigan, Minnesota, Oregon, the Dakotas, western Kansas and Nebraska, and those embedded in the Rockies and the Pacific Slope. California suffered major fires from Marin County to San Diego. Bakersfield burned. San Diego County burned. The Santiago fire, under howling Santa Anas, was estimated at over 300,000 acres. Flames went "to the very gates of Portland." The fires in Yellowstone National Park offered the first serious challenge to the U.S. Cavalry as a fire control agency (it failed, if gallantly). Elsewhere—all the elsewheres of the region—the fires flared and bolted. Place after place found itself under a pall of smoke, with ash and cinders drizzling from the sky like snow. In the Lake States the fires burned in May and again in October. In the Northern Rockies they burned as widely as the 1910 fires; their far-ranging smoke smothered Denver. But those shouting flames had no echo from the culture. They were part of the ecological background noise of the Northern Rockies. That didn't change until 1910.

As a natural phenomenon only the regional scale of the 1889 fire season distinguished it. The worst outbreaks occurred where the land was most disturbed by mining, railroading, logging, and settlement generally. The causes were the usual suspects: lightning, railroads, prospectors, brush-clearing farmers, sport hunters, and the always-suspect transients. (In southern Oregon "tramps," if not treated as well as they wished, were said to set revenge fires. Outside McLean's three were caught and lynched, but cut down and badly beaten before they died.)[2] On the Dakotas prairie fires were reportedly set by gangs collecting the bleached bones of buffalo to sell for fertilizer. There was no prevention program.

There were no agencies committed to controlling fires before they brushed against towns and mills. Then companies and city folk would rally and send hundreds of men out to hold the fires back.

There were no effective institutions for fire at all. Both Idaho and Montana were still territories. Montana's statehood convention met in Helena while "great black clouds of smoke hang over the country, and for six days the sun has not been able to pierce the darkness."[3] Two months after snow and rain extinguished the fires, it was granted statehood. Idaho became a state the next summer. Until then it was overseen by appointed rather than elected officials. Idaho's Governor Shoup telegraphed Secretary of the Interior John Noble for aid to keep fires out of Boise, and received $500. After fires ringed Ketchum, Secretary Noble authorized another $500.

It was laughable, of course, but there were no mechanisms by which to apply any funds. Revealingly, that same year the American Forestry Congress convened and campaigned for a system of forest reserves and professionals to administer them. They looked of course to Europe, but also to Europe's colonies. The *New York Times* reported a rhetorical question asked by many at the congress: "Why should we not take the same precautions for protection against fire that are taken in India?"[4] That would require government bureaus. The creation of forest reserves was two years in the future. An organic act for the reserves was eight years away. The creation of the modern Forest Service was 16.

There were few mechanisms for reporting on the fires. No one added up the area burned, the men on firelines, the dollars spent or lost to damages. No one really knew where the fires were other than they were burning over the "whole county" or "whole countryside." Most reports came from random dispatches delivered to random newspapers. The most complete accounts on the California fires appeared in the *Lawrence Daily Journal* from Lawrence, Kansas. The reports on Secretary Noble's outlays for firefighting were reported in the *Bismarck Weekly Tribune*. An extensive survey of Montana fires, both prairie and mountain, came out of the *Fisherman and Farmer* in Edenton, North Carolina. The story of Portland's battle was published in the *Roscommon Constitutionalist* in Boyle, Roscommon County, Ireland.

The Missoula *Daily Gazette* might declare, as it did on August 16, that "from the northernmost boundary of British Columbia to the Mexican line, the broken mountain range is all ablaze . . . . The latter part of the dry season upon the Pacific Slope is always marked by forest fires more or less extensive. The entire absence of rain over long periods presents a predisposing condition. This year the destruction is unparalleled." They could at least see smoke through their windows and walk on the ash that sifted down on the city's streets. The *New York Times* had to rely on the report of a "gentleman just in from Northern Montana" as its source. A fire in the Adirondacks was in the *Times*'s backyard. A fire in the Rockies might as well be across the Pacific. (In fact, a fire reportedly in Suchow, China, received as many column inches.) For nearly everyone the 1889 fires that mattered in the greater Northwest were those that burned Spokane and Seattle. However much big burns might rattle the region, they failed to register on the nation's cultural seismographs. They were a snag that burned and fell in the woods with no one to hear it.

In retrospect it surprises, though it shouldn't, how much the 1889 fire season fits the template that prevailed in the Northern Rockies between the advent of mining rushes in the 1860s and the onset of the national forests in the 1890s. The same season of fire arrivals and departures. The same fire causes. The same droughts and high winds. The same fire rhythm of creeping and sweeping. The hastily assembled crews of miners, loggers, cowboys, railroad gangs, and vagrants. The towns circled by flames. The isolated cabins burned over. The stories of desperate flights, refuge sought in mine shafts, trains laden with evacuees crawling through dense smoke and flame. The whispered rumors of isolated deaths. The same reporting, at once breathless, mundane, and macabre that could speak with one breath about whole mountains ablaze and whole communities arraigned to struggle against them, suns blotted out by smoke, scores of miners missing, a Swedish family maimed as they fled, and "telegraph lines somewhat demoralized by burning poles." The pieces were interchangeable with big fire years across decades.[5]

Previously, before the arrival of westering Americans, another set-piece encounter had prevailed. And later, after institutions were created to preserve and manage a public domain, yet another template for big-fire narratives came into play. Throughout, big fires persisted. Undoubtedly they changed as they feasted on the slash of colonizing Americans, threatening a timber famine; then on wildlands made wilderness, overgrown with combustibles by being denied fire; and then as the exurbanites recolonized the rural scene and as climate inflected into a new normal. With each phase change, the meaning of those big fires for their sustaining society morphed, along with how they would interpret the big burns of the past.

The fact is, outside of devoted fire researchers and the occasional antiquarian, no one has much cared about the 1889 season. At best the 1889 fires were a dress rehearsal for the 1908 fires, which were a prelude to the Great Fires of 1910—and these were the fires that mattered. They were not the largest, or powered by the most horrendous drought or the highest winds, but they are the ones that burned with the greatest severity for American society. They established, for the first time, the terms of the American way of fire protection in wildlands. They traumatized the agency—the U.S. Forest Service—that more than any other laid down the infrastructure for nationalizing their lessons. They inspired more archives and gathered more records than any wildland fire previously. They wrote the creation story. We are still living with their aftershocks. We aren't with those of 1889.

═══════════

Wallace Stegner once observed that "no place is truly a place until it has had that human attention that at its highest reach we call poetry." That's as true for fires as for waterfalls, and mountain peaks, and sandstone gorges. A compilation of big fires is little more than a scatter diagram of history; without context it is a data set, not a narrative. Its Cartesian points are so many snags that burned and fell in a forest and no one heard, at least in a way that anyone else cared about. The Northern Rockies fire scene has repeatedly conjured up enough big fires with a rough approximation of backwoods poetry to influence national policy and define a Platonic ideal for what fire protection in the form of a set-piece firefight might look like. Yet there is a problem with the kind of meaning those big fires created.[6]

The American fire community is not prone to citing William James. But as they seek to turn fires into data, and data into narratives of meaning, they might consider one of James's essays, "What Makes a Life Significant?" Here James argued that what animated events, and lives, was conflict and struggle—a desperate contest, uncertain in its outcome, pivoting on the will and strength of character of the actors. This certainly holds for wildland fire as well, and it is a primary reason why the firefight has proved more compelling than the prescribed burn. The firefight against monster flames can serve as the Hollywood script of a desperate struggle, its outcome uncertain. The prescribed fire, by definition, is a controlled process whose conclusion has, in principle, been foreordained. A firefight has drama. A prescribed fire should have none.

No fire regime has only one kind of fire, and the Northern Rockies is no exception. But it has a defining fire. The big blowup in the backcountry shapes the region as much as Santa Ana–driven conflagrations do Southern California or near-annual surface burns do the Flint Hills. So, too, no regime if it is to be robust can depend on one story or triangulation pin. The challenge for the Northern Rockies will be to bond its new avatar of the big burn, the megafire, with cultural clout. In 1910 it did the trick by aligning with another classic essay from William James, "On the Moral Equivalent of War," which was published in *McClure's Magazine* the same month the fabled Big Burn roared over the Northern Rockies and blew through a Forest Service that still had the dew on it. The timing was mostly coincidental, though not entirely. The firefight as battlefield here found its philosophical apologist.

What the Northern Rockies will do in the coming era will determine how relevant the region remains to the national and even global saga of fire. A big fire can become a great fire only when the rhythms of blowups match the cadences of a colluding culture. If the Northern Rockies hold only with big smokes, bravura firefights, and young men lost to the flames, if it continues to recycle its memes as it does its big burns, pumping them into larger sizes without reconfiguring their meaning, it will fade into a historical palimpsest, as the once-burning Adirondacks and North Woods have. If, however, it can twist the prism of meaning to catch more fashionable themes, as for example with wilderness or sustainability, surely a desperate struggle of our times, it might refract those brilliant flames into new significance.

THE BIG BLOWUP

A meditation inspired by the rededication of a memorial at St. Maries, Idaho, on the occasion of the centennial of the Big Blowup.

And upon earth he shewed thee his great fire;
and thou heardest his words out of the midst of the fire.
DEUTERONOMY 4:36

A HUNDRED YEARS AGO, on this day, at this spot, American society and American nature collided with almost tectonic force. Spark, fuel, and wind merged violently and overran 2.6 million acres of dense and odd-disturbed forest from the Selways to the Canadian border. The sparks came from locomotives, settlers, hobo "floaters," backfiring crews, and lightning. The fuel lay in heaps alongside the newly hewn Milwaukee Railway over the Bitterroots and down the St. Joe valley, and across hillsides ripped by mines and logging, and still-vast untouched woods primed by drought. The Rockies had experienced a wet winter but a dry spring that ratcheted, day by day, into a droughty summer, the worst in memory. Duff and canopies that normally wouldn't burn now could. The winds came with the passage of shallow cold fronts, rushing ahead from central Washington and the Palouse and the deserts of eastern Oregon, acting like an enormous bellows that turned valleys into furnaces and sidecanyons into chimneys. Southwesterly winds rose throughout the day to gale force by early evening, and then shifted to the northwest. Perhaps 75 percent of the total burn occurred during a single 36-hour period, what became known as the Big Blowup.

This was the first great firefight by the U.S. Forest Service, or for that matter, by the federal government, which is to say, the nation as a whole. Large fires had followed frontier landclearing like rats. But these fires were on public wildlands, and they became the first contest that challenged national power and will. As the scene evolved it acquired all the parts that would define the American way of wildland fire. All the pieces slammed into place for the first time. The fires called into being the whole apparatus from a search for crews to emergency funding. Equally, it defined their meaning. It established the discourse between fire lighting and fire fighting. It crafted a story to explain what had happened and what it signified. It established a narrative.

As the weeks wore on, the fires had crept and swept, thickening during calms into smoke as dense as pea-soup fog, then flaring into wild rushes through the crowns. The fledgling Forest Service, barely five years old, tried to match them. It rounded up whatever men it could beg, borrow, or buy and shipped them into the backcountry. The regular army contributed another 33 companies. The crews established camps, cut firelines along ridgetops, and backfired. Over and again, one refrain after another, the saga continued of fires contained, of fires escaping, of new fronts laid down. Then the Big Blowup shredded it all. Smoke billowed up in columns dense as volcanic blasts, the fire's convection sucked in air from all sides, snapping off mature larch and white pine like matchsticks, spawning fire whirls like miniature tornadoes, flinging sparks like a sandstorm. Crews dropped their saws and mattocks and fled. That day 78 firefighters died. One crew on the Cabinet National Forest lost four men; one on the Pend Oreille lost two; the rest of the dead fell on the Coeur d'Alene.

The Coeur d'Alene was ground zero. In the St. Joe Mountains between Wallace and Avery, some 1,800 firefighters and two companies of the 25th Infantry manned the lines when the Blowup struck. A crew north of Avery survived when Ranger William Rock led them to a previously burned area, except for one man who, panicking, shot himself twice rather than face the flames. A crew on Stevens Peak lit an escape fire in bear grass, then lost it when the winds veered, and one man died when he stood up and breathed the searing air. A crew at the Bullion Mine split, the larger party finding its way into a side adit; the rest, eight in all, died in the main shaft. On Setzer Creek some 28 men, four never identified even as to name, perished as they fled and fought their way uphill and fell in a collapsing ring of death. A gaggle of 19 spilled off the ridge overlooking Big Creek and sought refuge in the Dittman cabin. When the roof caught fire, they ran out. The first 18 died where they fell, in a heap along with five horses and two bears; the 19th twisted his ankle in crossing the threshold and collapsed to the ground, where he found a sheath of fresh air. Two days later Peter Kinsley crawled, alive, out of a creek. Another group dashed to the Beauchamp cabin, where they met a party of homesteaders. A white pine thundered to the ground and crushed two men immediately, while trapping a third by his ankle; he died, screaming, in the flames. Another seven squirmed into a root cellar where they roasted alive.

And then there was the crew cobbled together by Ranger Ed Pulaski. He had gone to Wallace for supplies and was returning on the morning of the 20th when the winds picked up their tempo and cast flame before them. He began to meet stragglers and then a large gang spalled off from the main ridge camp. All in all he gathered 45 men, and with the smoke thickening in stygian darkness turned to race down the ravine of the West Fork toward Wallace. One man lagged and died in the flames. Pulaski hustled the rest over the trail before tucking them into a mine shaft. Then he hurried down canyon with a wet gunnysack over his head before returning and herding the group into a larger tunnel, the Nicholson adit, which had a seep running through it. Pulaski tried to hold the flames out of the entry timbers and the smoke out of the mine with hatfuls of water and blankets. But by now the men were senseless. They heard nothing but the din, felt nothing but heat, saw nothing but flame and darkness, smelled only smoke and sweat. As the firestorm swirled by the entrance, someone yelled that he, at least, was getting out. At the entry, rudely silhouetted by flames, he met Ed Pulaski, pistol drawn, threatening to shoot the first man who tried to flee.

Those were the rude events. They ended with 78 firefighters dead, a sickened but enduring Ed Pulaski, an agency and a landscape deeply traumatized, an institution hopelessly indebted, and a policy under fierce scrutiny. From its origins the Big Blowup had been a double firefight, one in the field against flames and one in Washington over the politics of fire and state-sponsored conservation. But it is just such entanglements with a broader culture that make big fires into great ones. The 1910 fires swept over the Canadian Rockies with hardly a whisper, but they smashed against the American Rockies because here the hammering flames struck an institutional anvil with a clang that still rings through America's wildlands.

The Great Fires of 1910 seemed decisive in their significance—how could they not be? But in truth a long struggle emerged to extract meaning from the events, and it is that ambivalent ordeal that brings us here today. We are commemorating not so much the facts—there have been fires larger, more expensive, more damaging, more starkly flung in the face of public opinion, and more deadly; we are remembering the significance that has been attached to those facts. The Great Fires became the forge for a century of American fire policy and practice. They became our national fires of reference. Commemorations do not emerge of themselves. They are made, and how they are made is a part of what they celebrate.

It took many decades, but what happened that day evolved into a creation story. This was a struggle as troublesome as manning firelines, for the meaning of the fires could veer as unpredictably and suddenly as the flames on the St. Joe Mountains. The belief that fire ought to be managed on the public lands was, in 1910, an idea with many prophets but few converts, and the assertion that such fires ought to be fought was even more heretical, and probably flaky. Most of the general public was indifferent or hostile to aggressive fire control, bar fires that immediately threatened property or lives. Rural Americans relied on fire—burned everything from ditches to fallow fields—and accepted the occasional wildfire as they did floods or tornadoes.

So as reports screamed across telegraph lines, it was not clear how the fires would be interpreted. Those actually on the ground considered the Great Firefight an utter rout. On the Lolo National Forest, supervisor Elers Koch declared the summer a "complete failure." Despite unparalleled efforts by the Forest Service, assisted by the regular army, the flames had roared over the Bitterroots with no more pause than the Clarks Fork over a boulder. At national headquarters, foresters fretted whether the Great Fires might be the funeral pyre of the besieged Forest Service. In fact, those far removed from the flames saw them otherwise. They chose to see Pulaski's stand, not his flight. They saw a gallant gesture, not an act of desperation. The Forest Service's critics claimed the service had been granted ample resources and had failed. Its defenders replied the Service failed only because it had not been given enough. Both groups could point to the Big Blowup for empirical support.

Perhaps surprisingly, the political tide turned in favor of the Forest Service. The agency successfully defended its 1911 budget. The Weeks Act that would provide for the eastern expansion of the national forests by purchase and for federal-state cooperative programs in fire control, stalled for years, broke through the congressional logjam in February. In March a beleaguered Richard Ballinger, Gifford Pinchot's political rival, asked to resign. Foresters redoubled their efforts to crush light burning, and all it implied; then they turned on other scoffers of fire protection. By now the Forest Service had the memory of the fires spliced into its institutional genes.

The Great Fires were the first major crisis faced by Henry Graves, Pinchot's handpicked successor. The next three chief foresters—William Greeley, Robert Stuart, and Ferdinand Augustus Silcox—were all personally on the scene of the fires, had counted its costs, buried its dead, seized upon "smoke in the woods" as their yardstick of progress. Not until this entire generation passed from the scene would the Forest Service consider fire as fit for anything save suppression. Three months after the Big Blowup, Gus Silcox wrote that the lesson of the fires was that they were wholly preventable. All it took was more money, more men, more trails, more will.

In 1935 Silcox, then chief, had the opportunity to reconsider. The Selway fires of the previous summer had sparked a review in which the Forest Service itself admitted the lands it was protecting at such cost were in worse shape than when the agency had assumed control. Field critics observed that the service was unable to contain backcountry burning. Scientific critics had announced at the January meeting of the Society of American Foresters that fire was useful and perhaps essential to the silviculture of the longleaf pine. Ed Komarek observed bitterly that this was the first time such facts had become public. And a cultural criticism burst forth as well. Elers Koch noted that the pursuit of fire into the hinterlands—mostly by roads—was destroying some of the cultural value of those lands. The Lolo Pass, through which Lewis and Clark had breached the Rockies, he lamented, was no more, bulldozed into a highway. All this landed on Silcox's desk. His reply was to promulgate, in

April, the 10 a.m. policy, which stipulated as a national goal that every fire should be controlled by ten o'clock the morning following its report. The veteran of 1910 replied, that is, by attempting to squash fire, to allow it no sanctuary, to tolerate no qualifications, to apply the full force of the Civilian Conservation Corps and the federal treasury. He would refight the Great Fires, and this time he would win.

In this way Chief Forester Silcox identified the "lesson" of 1910 and applied it in 1935. But history is full of lessons—it overflows with them, its landscapes are littered with lessons like jackstrawed lodgepole. The issue was not whether there were lessons but which lessons to seize upon, how to interpret their particular significance, and what parts of them to apply in very different circumstances. The Great Fires were as ambiguous as the blowing up of the battleship *Maine* in Havana Harbor or the 1938 concessions at Munich. It meant something, and something that should not be repeated. But what exactly, and what that might mean in another time, was unclear and subject to reinterpretation. Like post–Civil War Republicans waving "the bloody red shirt" it could mean everything, and nothing. It was not certain that the Great Fires mattered beyond a founding generation of foresters or outside the Northern Rockies. For most issues the American fire community has a notoriously short memory. It obsesses over next year's firefight, not last year's, much less last century's.

The construction of meaning proved as arduous as, if less lethal than, the trenching of firelines. The painful inability of the agency to find a suitable memorial for its fallen firefighters testifies both to the intensity of the fire's shock and the agency's fumbling. No one had imagined scores of dead, some of them never identified even as to name; there simply existed no bureaucratic mechanism to inter them properly, much less to honor their sacrifice. And no one had expected a similar need to translate those brutal facts, buried or otherwise, into a narrative that could endow them with the kind of meaning that makes for a usable past.

It came about slowly. It began with written records, which are unusually rich. Forest Service supervisors wrote narratives. Claimants for compensation, allowed by a special act of Congress, recorded their experiences. Local newspapers were full of accounts. Some participants wrote

and cached letters or diaries. The army added its own chronicle of tele-
grams and clipped reports. Remarkably, an agency photographer even
toured the scene a few weeks afterwards and bequeathed a visual record
unrivaled in American fire history. For a while, during the 1920s, Region
One collected memories from still-living survivors, but was unable to
organize them into a coherent narrative. In 1943 Elers Koch gathered
the files together into an anthology, *When the Mountains Roared: Stories
of the 1910 Fire*, subsequently published by Forest Service mimeograph
machine. In the 1950s Betty Spencer organized these materials into
a rough narrative, published with a regional press as *The Big Blowup*.
That led to interest in a commemoration for the 50th anniversary of the
Great Fires.

The 50th anniversary contrasts sharply with today's. While fire
remained fundamental to the agency, it had evolved into a mild annual
ritual or annoyance rather than an extraordinary catastrophe that threatened to burn the agency to its roots. In the summer of 1910 Henry Graves
had declared that fire protection was 90 percent of American forestry. By
1960 it commanded about 10 percent of the agency's operating budget.
The Forest Service had become a fire hegemon. Such as there was, interest in commemorating the Great Fires was sparse and resided in local
personalities. Knowledge of the crisis had so ebbed that no one could
locate the site of the Nicholson adit, better known as the Pulaski tunnel,
the defining episode of the Blowup. The memorial was strictly a regional,
even parochial, event.

Today, the fires' commemoration has achieved a wider reach. The reasons are many but among them are the revival of wildland fire as a national
spectacle and political problem, the replacement of a fire-threatened
rural frontier with a fire-threatened exurban one, and a reconnection of
fire with the broader culture, for which Norman Maclean and *Young Men
and Fire* are largely responsible. By restoring its valence to the national
culture, the Big Blowup has reclaimed its status as a great fire. The Nicholson adit has been declared a national historic site. The Pulaski tunnel
now climaxes an interpretive trail. The Pulaski tool is universally identified as the distinctive symbol of wildland fire management. The Pulaski
story has become the plot pivot to our national narrative of fire. That
is why we are here today, and why the principal sites of the fires have
become hallowed ground.

Today the narrative legacy of the Great Fires reads as follows: The young Forest Service found itself threatened on many fronts, and fire was not only among those threats but a visible test on its ability to do what it claimed needed doing. After the Big Blowup it determined to exclude fire so far as possible. It unwisely took out of the system a necessary natural element and created an institutional juggernaut that seemed ready to destroy the forest in order to save it. For the past four decades the federal agencies have tried to pick up the pieces from the wreckage wrought by fire control, much as they sought to sweep up after the Great Fires. The buildup of fuels and the havoc of fire-famished ecosystems have spawned megafires and degraded biotas. Today's need is for more of the right kind of fire. The master narrative ends with a modernist twist, the implied irony that by honoring the victims of the Big Blowup we are perversely celebrating a massively wrong turn in American environmental history.

That is where that narrative must end. But I don't believe that is where we want it to go. The narrative of the Great Fires explains much of why the scene today looks the way it does. It does not explain what it should look like. It does not tell us what we should do. On the contrary, it suggests that we might simply stand aside and let nature return fire on its own terms. Those burned landscapes, however, were not the result of natural processes alone, and a narrative that presents the informing conflict as one between people and nature cannot inspire the invention of a new landscape. For that we need a new narrative, one that does not begin in August 1910 nor end in ironic condescension.

We don't need to scrap that old narrative—it's too good, and as we see about us, it may be indispensable. The fires burned in a moral universe no less than in the mountains. Their story reminds us that we live in a world about which we know only a little yet a world in which we must act, which is to say, in a world for which character—hubris, recklessness, egotism, tenacity, courage, and loyalty—matters more than rates of spread and fireline intensity. So even as we retell the epic, we should recognize that we need to fashion another story, one equally compelling and convincing, to add to it. Again, the Great Fires can instruct, for they tell us what it takes to embed a fire in a culture in ways that create a usable narrative—a narrative that can inform and inspire. Fires alone, no matter how large, are not enough. It is how they engage with the cultural ecology of their times that matters, and few of the essential ingredients are under our control. That is what makes narrative history different from novels and why narratives that emerge from lived experience are so difficult.

It took years, then decades, and now nearly a century to create suitable memorials to those who fought the fires. These sites bear witness to how the Great Fires of 1910 engaged with Progressive Era America. The story of how Americans and their fires interact today has not yet gelled into comparable symbolism. A century hence, let us hope we will have crafted such a narrative, a testimonial better suited to the complex relationship we have with our land, one not dominated by metaphors of fighting and dying but of working with and regenerating.

Today we can honor those who fell in answering a call to arms and adventure. Tomorrow we can hope such calls will indeed be a thing of the past.

THE SECOND BIG BLOWUP

The Almost Great Fires of 1988

EVERYONE AGREED AT THE TIME that the fires of 1988 were a monumental event, not only for the Northern Rockies but for the nation; and that judgment has persisted. For years afterward the big burns were the pivot of conferences, training sessions, scientific studies, and fire cache chatter. Twenty years later they merited a retrospective conference. The fires were big news. They seemed a graphic demonstration of the fire revolution. They appeared to underscore a phase change in the relationship among Americans, their lands, and their fires. They were epochal.[1]

But it was never clear what exactly they signified and why they mattered. Were they important because they changed American policy and practice, and made visible those reforms because they occurred at Yellowstone? Or were they a celebrity event, made important because they happened at a landscape with global cachet? Are they known today because they reshaped America's fire institutions as they did lodgepole pine around the Yellowstone Lake? Or because, like celebrity itself, they are known for being known?

———

A box score of the fire season more or less speaks for itself. The region experienced 4,168 fires that burned 2,175,903 acres. On September 10 some 15,700 personnel were deployed against them, roughly 9,500 at Yellowstone.

Air operations racked up 24,950 hours of flight time. Within the Greater Yellowstone Area (GYA) 1,061,995 acres lay within fire perimeters, of which maybe 30 to 40 percent remained unburned. Estimated costs for the GYA fires begin at $120 million. Media coverage ran to untold numbers of words and film footage; by the end of July the fires had become a fixture of the national news, and remain so. Before the Yellowstone blowup a big western fire might get noticed typically for a night or two. Yellowstone being Yellowstone, fire coverage went on night after night after night for over six weeks, the longest-running televised serial in American fire history. On Black Saturday, August 20, the 78th anniversary of the Big Blowup, the fires doubled the area burned. The climax came on September 7, when flames washed over Old Faithful and threatened developments laid down from the days when the U.S. Cavalry ran the park.[2]

The symbolism was overpowering, but it was also oddly indeterminate, as shape shifting as the flames. It just was. The bust was destined to be politicized, an outcome worsened by an election year and photo ops. Out of its ashes reviews sprouted like morel mushrooms. Within the fire community discussion went on and on, rather like the newscasts, but for years. Unquestionably, the 1988 fires—the Yellowstone fires, as they became known in shorthand—were a big event. By most standards they were the biggest since 1910. To some minds they constituted a second Big Blowup.

In size and shock value that is likely true. And like the Great Fires of 1910 those of 1988 overwhelmed the system as the need to act blew over plans, ideals, preparations, and existing knowledge. Expert opinions about how the fire season would evolve and the probable final acreage burned went up in convective plumes unlike any the old fire dogs and new computer-equipped wizards had ever seen or could imagine. (On August 1 Dick Rothermel predicted that 200,000 acres would burn in the GYA. Don Despain announced that the fires would soon run out of fuel. They were off by a factor of five.) Nor could fire agencies any more control raging public opinions and social consequences than they could the flames, although they rallied around both the notion of fire's restoration and the valor of the firefight that followed. After the Big Blowup, the dissenters were few. While Elers Koch might despair that the Forest Service had suffered a rout in 1910, most rangers and supporters of government-sponsored conservation followed the lead of Gus Silcox who argued that with more resources and better public support fire suppression could be made to work. So, after the Blowup of 1988, skeptics were brushed aside as an ill-informed sect. With renewed commitment the larger project of the fire revolution could succeed. The Yellowstone conflagration became, in fact, the squeaky wheel that brought a lot of grease to the National Park Service.[3]

Enough time has passed that a more textured historical comparison is possible. After all, the outcomes to big events are known by their contexts, and those settings may evolve over several decades. It took 25 years for the major aftershocks of the Big Blowup to make themselves felt. As we pass the 25th anniversary of the Yellowstone burns, it's a good time to survey the setting for what the fires did and haven't done. While a too-close reading hedges into historical astrology, consider the pre- and post-fire chronologies as parallel texts amenable to benign glossing.

A scan might look like this. The national forest system began in 1891, and received an organic act in 1897. The Big Blowup occurred, respectively, 19 and 13 years later. The National Park Service adopted the Green Book for administering its natural areas in 1968, and Yellowstone fashioned its new-order fire plan in 1972, or 20 and 16 years, respectively, before the 1988 blowout. A year after the Great Fires of 1910, Congress passed the Weeks Act, which created the federal-state infrastructure for fire protection and Coert duBois published the precepts that would underlie *Systematic Fire Protection in the California Forests*, which in turn guided universal planning for fire protection. In the year following Yellowstone's burns, an interagency committee both reaffirmed and rechartered federal fire policy, after which all of the public domain lands with natural fire programs had to reboot according to the new software over the next few years. The upshot of the Big Blowup, the 10 a.m. policy, was promulgated in 1935, or 25 years later. The National Cohesive Strategy, intended to shape national fire policy, was scheduled for 2013, or 25 years after Yellowstone. If you want to argue that the Yellowstone fires had a catalytic impact comparable to the Great Fires of 1910, you can make a case.

Still, you don't need allusions and comparative scenarios to identify the Yellowstone bust as a major moment in American fire history. The bust led directly to a review of federal policy, and brought fire into the radar screen of the Government Accountability Office. The evolution of a common federal policy as distinct from a collection of reforms among individual agencies slowly congealed out of the inchoate ashes of the North Fork, Clover-Mist, and Wolf Lake complexes. To upgrade capabilities the National Park Service, in particular, received a boost in funds and attention that helped propel it into the vanguard of fire programs for the next decade. The Yellowstone fires diverted attention from the Canyon Creek fire in the Bob Marshall Wilderness that might have compromised the natural fire program of the Forest Service, and so granted the USFS some space for maneuvering.

Mostly, the fires brought home to the American public, as probably nothing else could, the significance and practical consequences of the fire revolution. The reformation meant that iconic places like Yellowstone would burn—would need to burn. A broad understanding followed and largely took root, like the mass postfire reseeding by lodgepole pine. The message propagated nationally, and internationally. The Yellowstone example stood as a global exemplar, for good or ill. Much as its convective columns spewed ash downwind, the fires sent images and information around the Earth through the plumes of TV, popular print, and technical journals.

Within a year it was clear that the revolution had withstood the test. Its message, if not the particulars of its policy, had passed through the flames. Twenty years later that conviction—that Yellowstone mattered—warranted a convocation of those concerned under the rubric of a Tall Timbers conference. There is good reason to believe that by 2038 historians will judge the Big Blowup of 1988 as an event of major stature, if not quite comparable to that of 1910. Yellowstone's partisans, and they are many, would demand nothing less.

And yet the second Big Blowup was a lesser echo. The Great Fires of 1988 did not splice themselves into institutional DNA as those of 1910 did. When the Big Blowup struck, the Forest Service was an adolescent agency composed, as one member put it, almost wholly of young men. When the 1988 fires hit, Yellowstone had been a park for 116 years, had known fire protection for 102, and was overseen by an agency 72 years old. Nor was the NPS positioned in 1988 to influence national policy as the Forest Service was after 1910. The Yellowstone fires did not reform policy. They did not instruct the fire community beyond the fact that it did not know as much about extreme fire behavior as it thought and needed to engage the public better in its deliberations. They did not redirect the course of American fire history. Efforts went into defending the program, holding to what had been gained, not advancing it. Still, if the NPS could not use the Yellowstone fires to promote the future, it did not cave in to demands from the past. The firelines on the ground failed. Those in the mind, and in policy, held.

The big difference is that the Great Fires of 1988 did not have an Ed Pulaski to symbolize its terrors and resolve, or to codify its lessons in a tool. The Big Blowup invented the narrative of the wildland firefight, and adapted an inherited narrative of the conflagration as a disaster story. While commentators tried to apply both to Yellowstone, neither fit. Apologists for the park and for natural fire refused (rightly) to allow the disaster template to apply. The firefight narrative faltered because, despite an immense investment in personnel and equipment, it failed, save for shielding some structures. The grand difference between 1910 and 1988, in brief, is that the Big Blowup created an enduring narrative and the Big Blowout did not. It left the interpretation of the fires—so vast they just had to mean something—unresolved.

Except for that last thought, this quick panorama encapsulates the received standard wisdom of the American fire community. For a long time, however, I have resisted and withheld agreement. My reasons are personal. They stem from the summer of 1985 when, at the request of the Rocky Mountain regional office of the NPS, I wrote a draft fire plan for Yellowstone.[4]

Over the two previous summers I had worked at Rocky Mountain National Park as a fire planner. I thought I was headed to Theodore Roosevelt National Park for 1985 when Jim Olson redirected me to Yellowstone. The park's program did not meet national guidelines as codified in NPS-18, did not align with neighbors, and not a few observers outside Yellowstone regarded the place as an accident waiting to happen. Although Yellowstone National Park had size and glamor, it did not have an operational fire plan. I was there to coax one into being.

I found a park staffed with dedicated long-termers (homesteaders, in NPS parlance), a group tenaciously committed to the Yellowstone ideal as they understood it. They were also remarkably insular and self-referential and indifferent or hostile to outsiders. For that last sentiment I could hardly blame them. Still, Yellowstone did not have a fire plan; since 1972 it had a statement of philosophy that sought to encourage as much natural fire, which would mean crown fire, as fast as possible. (It was also clear that what underwrote the fire plan—and at times, everything else—was the enduring conundrum of elk.) My task was to translate that sentiment into formal language. I spent 10 weeks researching, exploring, and writing, and produced a set of documents, which I presented to Superintendent Bob Barbee and the park's Fire Committee.

The final package had three items. One was a fire plan written according to the guidelines of NPS-18, the agency's reference manual for fire management. It offered a few novelties but mostly just filled in the blanks. The second was a slate of "recommendations for future actions." The third was a prospectus for a Yellowstone Interagency Fire Management Center that would ground the park's ambition to restore natural fire in an institution and make the Greater Yellowstone Area the centerpiece for natural fire everywhere. "It will seem odd for an observer to insist that Yellowstone take itself more seriously, rather than less so, but the opportunities for fire management are special, and Yellowstone, with all humility, should assume the burdens—not merely the status—of leadership." The prospectus went nowhere (it had no champion, and so no chance). The recommendations gathered dust. The fire plan was hefty and formal enough to get the regional office and Branch of Fire Management off Yellowstone's case, though the plan was never submitted for public review or even vetted through the agency. The park quietly excised all the checks and balances on the Fire Committee that the draft spelled out. None of the plan's protocols that I had inserted per NPS-18 were followed during the 1988 season. I was not surprised by the outcomes, though I underestimated the maximum fire year by a factor of two. But my experience made me observe the season through a glass, darkly.

My objections were two. I thought the park, its apologists, and the American fire community overall missed a once-in-a-lifetime opportunity to present the fire revolution to the American public in a way that went beyond name-calling and slogans. Instead, partisans fixated on the wrong question, which is to say, whether fire belonged in Yellowstone. That was easy: of course it belonged. Initially the media recycled the usual clichés about "destroyed" landscapes and the like, and politicians railed about ineptitude and government waste (it was an election year); but rather quickly that blather disappeared. The educable public learned and accepted. The fires were not an ecological catastrophe. To my calculations they burned off in one year what would have burned over the past hundred had the park and army-inspired fire protection not intervened. The official Yellowstone position was that firefighting had been ineffective against the crown fires that mattered. My reading was that the actions taken beginning with the cavalry in 1886 had stopped many ignitions before they became big and had virtually eliminated human ignitions altogether. Behind that distinction lay deeper assumptions about the ways Yellowstone was or was not a purely natural or a cultural landscape.

The brouhaha over fire ecology became an exercise in misdirection that prevented a much more important question from being asked, which was *how* does fire belong? at what cost? with what methods? according to what social compact? The Big Blowup of 1988 was not the only way to restore fire. The park, however, refused any other kind—was adamantly opposed to prescribed burning as a violation of the park's principles. Once the fires had swarmed over hundreds of thousands of acres nothing short of early snow could halt them. Whether or not the public was ready for that kind of discussion, the fire community should have been. In many ways, the community still yearns for it in a public forum.

And this was my second issue. The Yellowstone fire program, as I saw it, operated more as a personality cult. It had its own classification of fuels, apart from national standards. It dismissed National Fire Danger Rating System models. It relied, rather, on the highly personal knowledge of its particular members. The implied assumption was no one outside Yellowstone could know the Yellowstone scene, and Yellowstone knew all it needed to put fire back. It desperately wanted big fires. It was the flagship park, yet while others were actively accepting natural fires and kindling prescribed burns, Yellowstone's fire program was a script that wasn't being filmed. The only thing Yellowstone needed was ignition, drought, and wind: the park was so big that it could absorb whatever happened, both ecologically and politically. It didn't need protocols and prescriptions. That sentiment proved almost true.

A fire plan, as I saw it, was not a nuisance but a social compact. It was an agreement that specified what the park could do and not do, and would recognize the limits of any proposed action when confronted with what nature in its power and majesty could concoct. Yellowstone's view was that Yellowstone didn't heed rules; it made them. Plans were for little parks. The 1985 fire plan, as submitted, operated under the notion that ignitions would be handled as prescribed natural fires, which is to say, as events under prescriptions and within boundaries, and with procedures to evaluate and if necessary intervene. Those prescriptions were removed in the park's internal edit. Instead, the fires were let-burns.

I thought the result was cynical. It reeked of bad faith. If the park didn't like the constraints, it should protest and try to change them rather than ignore them, or use them as cover while it did what it wanted. I didn't like that outcome and didn't like that I had become by association an enabler of that deception and so didn't like the fires. I was wrong. I should have been on the scene. I should have self-deployed. Instead I stayed home and wrote a fire history of Australia.

Enough time has passed that the size of the event has shrunk relative to newer entries. Blowups have yielded to megafires. The 1994 season led to policy reforms. The 2000 season established itself as the natural successor to 1910. Some historical parallax is now possible, and with it, a few frames exist by which to measure and hang the event on history's wall.

I have made my peace with Yellowstone. Whatever unease my personal experience instilled, I have a duty to assess the Yellowstone burns as a historical event, and I can only conclude that they were a point of inflection. They were not the only tipping point, or even the most significant, but the history of American fire would be different if they had not occurred. Had Yellowstone wriggled out of its solipsism, had its partisans been more reflective, had the fire community been willing to criticize instead of instinctively circle its engines, had the fire been subject to thoughtful analysis by intellectuals other than scientists and journalists, Yellowstone might have become the hinge of the revolution. It might have given the revolution what it most needed, a narrative.

Instead, the era of wilderness fire succumbed to an era of the wildland-urban interface, a problem with little sense and less poetry. By its silence the fire community did not gain public acquiescence for the revolution. It denied the revolution a powerful story, and with it, the better part of a decade while reforms retrenched. Instead, the torch of reform passed to the 1994 season, in which there were fewer big fires but many deaths and a connection with the larger culture that went beyond celebrity. Yellowstone was big, but not, in the end, big enough.

THE OTHER BIG BURN

Reflections on Fire at Glacier National Park

G LACIER APPEARS LIKE A PATCH of First-Day nature still fresh from the Pleistocene. Everything about the place screams monumental, a vast sculpture garden from Nature's Romantic age. Its fundamentals seem far removed from human meddling. Snow, ice, streams, lakes; water hard, water soft, water flowing, water still; chiseling, caressing, abrading, covering—here water in all its forms acts on an upswelling of rock to make what most viewers would regard as an ur-scene of natural grandeur. Even amid the Rockies, the massif is striking, like a glacial erratic left from the ice age, moved not over space but through time. So immense are the domes and peaks that the cliff-covering forests appear like a veneer of lichens and those in valleys like moss. They frame the core scene like vignettes in an illuminated manuscript. Once the place was discovered, the founding question was never whether it deserved to be a national park, only when and how.

Yet Glacier's story is equally one of fire. The park was birthed amid the Big Blowup, its fires have repeatedly influenced the National Park Service, and its future may well be decided by the Big Burn of industrial combustion.[1]

―――――

When Major William R. Logan arrived to assume his post as superintendent of the newly authorized national park, Glacier was burning. It was an outlier of 1910's Great Fires, but the fires threatened the park's entry and core. The Forest Service assumed control, assisted by the Great Northern Railway. Perhaps 100,000 acres burned. Another 50,000 acres within the park burned in the regional fire bust of 1919.

Its fire regime has a northern rhythm. A big burn comes every decade or two, and the administering agency finds itself with too large a fire organization in the cool years and too small a staff in the hot ones. What has aggravated the scene is the feudal nature of the national parks, which is less a system than a coalition of baronies, each park separately created and governed. Not until 1916 did the country create a civilian agency, the National Park Service, to assume responsibility from the U.S. Army. With its scattered holdings, the NPS found it hard to match fires and staffing. There were few buffers and backups; unlike the Forest Service, it did not have access to emergency funding for firefighting; and the administrative apparatus could crack when its dispersed political structure, which could not muster much collective response, met a northern fire economy in which the extreme years drove the system. In this way Glacier both miniaturized and exaggerated the national conundrum. As a big fire could upend Glacier, so a big-fire year at Glacier could unbalance the National Park Service. Just that happened in 1926. In 1910 the park had essentially no staff when the fires began; in 1919, it was still a fledgling, finding its wings as a civilian agency; but in 1926 Glacier was a relatively mature park and one of the system's crown jewels. Several fires began outside the boundary, and one burned across the western border on July 7. Public complaints led the Department of Interior to dispatch Horace Albright, then assistant director of the service, to handle them. Then on July 31 a fire started from an exploding gas tank on a salvage logging operation within the park that swelled around Howe Ridge into the Lake McDonald fire, and high winds on August 5 then sent another fire, the West Huckleberry, over the north slopes of Apgar Mountain and into the Lake McDonald burn. The two fires merged, blew up, burned over 50,000 acres, and cost $230,000 to fight (nearly $3 million in 2012 dollars). Without emergency funds to draft from, the firefight caused the Park Service to shutter some parks to pay the bills. The 1926 fire season was, by almost any measure, a disaster—financially, administratively, politically.

The big burn worked on the NPS much as the Big Blowup had the Forest Service. It made a local issue into a national one, a move boosted the next year by the establishment of the Forest Protection Board to oversee the federal government's commitments on forested land and fire protection. The Board required fire plans, of which the NPS had none. Planning demanded someone knowledgeable to write plans and administer them. The service recruited John D. Coffman from the Mendocino National Forest, a man toughened by fights with local ranchers over light burning and familiar with Coert duBois's concepts of systematic fire protection, which had been devised for California. He focused on a national scheme to satisfy the Forest Protection Board with a report released in 1928, and the creation of a fire organization at Glacier to prevent a repeat of the 1926 debacle and furnish an exemplar for the system overall. The outcome could do for the national parks what duBois's treatise had done for the national forests. In early 1929 he and chief ranger F. L. Carter wrote that fire plan.

Still, in 1929, as Albright became director, the NPS's national fire organization amounted to Coffman, a solitary fire guard at Sequoia National Park, and the planned operation at Glacier. By August fires had again invaded the park. The Half Moon fire, started outside the west boundary in logging slash, burned over 100,000 acres (50,000 in the park) and cost the service $120,000 to contain. Coffman again appealed to Forest Service techniques, this time a style of postfire analysis, to critique the Glacier fire program. The review became, like the park's fire plan, a model for the service.

Then the fires began to fade away. The fulcrum was the Civilian Conservation Corps, begun in 1933, which at Glacier fielded 1,278 enrollees organized into nine camps over the next eight years. The CCC erected much of the infrastructure of the park, including its fire program. Enrollees even removed thousands of acres of unsightly snags killed by the 1929 burn. The 1935 and 1936 drought years racked up most of the burned acres, 5,456 and 7,722 (the Heaven's Peak fire claimed 7,600), respectively; but together they amounted to a third of the area burned by the Half Moon fire. The 1936 outbreak destroyed park facilities at Many Glacier, led to mutual-aid agreements with the Blackfeet Indian Agency, and inspired a service-wide review of every park's fire program, the promulgation of national guidelines for protection, and accelerated training for CCC crews. In the 1940s, despite wartime drafts on manpower, the largest fire, the Curly Bear of 1945, burned only 289 acres. By the end of the war aerial fire control replaced the lost camps of the CCC. Smokejumpers were attacking fires by 1946; in 1953 aerial reconnaissance was supplanting all but a small handful of lookouts. Meanwhile, visitors started far fewer fires, and the national forests were better at holding fires before they could break over the boundary. The 1950s had only one bad year, 1958, when 33 fires accounted for 3,000 acres.

In 1963, as the Leopold Report introduced the fire revolution to the National Park Service, only 12.3 acres burned. Over the next years even that annual acreage fell by two orders of magnitude: 0.0 acres, 0.1 acres, 0.2 acres. Glacier National Park had become what Albright and Coffman intended—a model of fire's exclusion. When Glacier published a new fire plan in 1965, it retained the 10 a.m. policy as its lodestone. The park's residual fires seemed as much a relic of a former age as its mountain glaciers.

⸻

The illusion vanished in 1967. Big burns returned to the Rockies, and two of them slammed into Glacier. On August 11 drought and dry lightning kindled 20 fires; other storms ignited another 15. Two grew large, and both swept through lands that had burned in earlier conflagrations. The Flathead fire chewed through the Apgar and Huckleberry Mountains, the scene of the 1926 burn and, whipsawed by the wind shear of cold fronts, spilled outside the park boundaries. The Glacier Wall fire burned across the region of the 1936 Heaven's Peak fire. Between them the fires consumed 12,330 acres. The park threw everything it had at them.

Glacier was as well endowed for fire control as any unit of the National Park Service, but it could hardly cope with burns on a scale it hadn't seen for 30 years, and since the region was up in smoke with more than 30 fires over a thousand acres, there was little reserve it could call upon from allies. The park responded by hiring what labor and equipment it could (3,500 men); it brought in overhead teams including the best fire officers in the NPS, ordered helicopters and National Guard units, authorized whatever means were necessary, expended $2.5 million, and after the smoke cleared hosted an elaborate postfire review. The siege went on for 62 grueling days.

What prompted the review was political attention. Local inholders primarily gave Senator Burton Wheeler a club with which to beat the NPS, while the more sympathetic Senator Mike Mansfield had to respond to constituents as well. The Washington Office had to reply. The intent of the two-day review was to demonstrate that the Park Service was serious about fire suppression ("fire is today, without a doubt, the greatest threat against the scenic grandeur of our National Parks");

that it had done everything possible to beat back the two big burns; and that Glacier remained a crown jewel not only of scenery but of effective administration. A primary conclusion from Superintendent Keith Neilson was that the major failure lay in public relations, in not getting the proper information to the press and public.[2]

Yet through the course of the discussion there were queries, mostly a subtext, about the wisdom of sending in a phalanx of bulldozers, particularly against the Glacier Wall fire. Les Gunzel voiced the most emphatic concern. The Flathead fire threatened developed areas and private cabins and had been burned over at least twice in park history; the dozers were warranted. At Glacier Wall they only wreaked havoc, and left scars far more vicious than burned snags. He wanted explicit guidelines. Others, not in the firefight, looked from the Leopold Report to the costly wreckage and wondered why in the remoter areas the fires were being fought at all.[3]

The next year the service published its famous Green Book for the administration of natural areas, which replaced the 10 a.m. policy with options drawn from Leopold Report recommendations, including the need to restore fire through natural means where possible, through prescribed burning where otherwise necessary. The effects were felt most immediately in the Sierra parks, particularly Sequoia (a proponent of light-burning in its early years) and at Saguaro (where Gunzel was chief ranger). Yellowstone rewrote its fire plan in 1972. Glacier was slower; after all it had been a centerpiece of Park Service suppression for almost half a century or nearly all the existence of the NPS. The new policy nudged forward in 1974, and a revised fire plan in 1978 put forth some tentative feelers while still supporting suppression. In 1980 the fire organization drew a zone of 100,000 acres in upper elevations for natural fire; it was a safe strategy in part because there were no fires. From 1983 to 1984, in association with researchers at the Missoula lab, it experimented with a few prescribed fires. Other parks, many much smaller (and perhaps more nimble) or with fewer fire problems, were experimenting boldly with the new options. The era pointed to Wind Cave, Saguaro, even Rocky Mountain. But Glacier—a big status park with a history of big burns—was a laggard. When the 1988 season rolled over the Rockies, it attacked all starts. Even so, some 27,520 acres burned, the largest since 1936, although the Yellowstone fires mesmerized the national audience and deflected political attention away from Glacier's blowup as it did the

Canyon Creek fire in the Bob Marshall Wilderness, which shot out of the mountains and overran tens of thousands of acres.

This time serious reform followed from the postfire reviews not of Glacier's fire bust but of Yellowstone's. Every park with any aspiration of incorporating natural fire had to rewrite its fire plans. Glacier completed the task in 1991. It zoned patches for prescribed natural fires and conducted some low-complexity burns in Big Prairie, an isolated ponderosa pine savanna. But the push was on for more black: the program needed to show some burned acres. Then, as so often before, a fire in Glacier rippled throughout the system. The year was 1994, a season notorious mostly for running up the first billion-dollar suppression budget and for the catastrophic South Canyon fire that engulfed a crew outside Glenwood Springs, Colorado. Those events, as emergencies are wont to do, overshadowed at the time what happened at Glacier, which pivoted on a managed fire, not a crisis. By sucking attention like stars into a black hole, they granted Glacier and the NPS some political space.

The Howling fire began on June 23, 1994, from a lightning strike on the North Fork of the Flathead River near the park's western border. At the time Glacier was one of a handful of parks with the size, funding, and clout to allow prescribed natural fires. A large burn, the Starvation Creek fire, which burned across the Canadian border, was under the direction of an incident management team. Suppression resources were strained. To oversee the Howling prescribed natural fire the Park Service assembled an ad hoc team to predict the fire's behavior and plan for contingencies. The Howling fire did not blow up, and by late August it was blocked to the east by two other fires, both wildfires, but both controlled through a confinement strategy. The three fires burned together. After the Starvation Creek fire was controlled, the Howling fire overhead team assumed control for it as well. For 75 days the team stayed with the fire until responsibility was ceded to the park. The fire burned for 138 days.[4]

By accident, necessity, and daring the Howling fire experience demonstrated how to cope with long-duration fires by substituting fire behavior knowledge for heavy machinery, trading land for options to maneuver, and adapting opportunities presented by nature to the strategic purposes of the park. They began by making the best of an awkward situation. They ended by inventing a new mode of operation. The 1995 federal wildland fire management policy received its momentum from the shock of the South Canyon tragedy and the season's cost in lives, dollars, and burned area. But the outcome led to a steady liberalization of how to handle wildfires that dissolved the prescribed natural fire in favor of the wildland fire use or resource benefit fires. Out of the Howling fire team emerged the idea behind fire use managers and modules. To the architects of those ambitions the Howling fire was the proof-of-concept test. Once again, without drawing national attention, Glacier had helped reform how the country would tend its fires.

Still, not until Glacier amended its fire management plan in 1998 did the park fully leap into the new era. Within a year it was managing the large Anaconda fire as a wildland fire-use burn. During the 2000 season, which ravaged so much of the Northern Rockies, Glacier suppressed the fires it received, but selectively, letting many blow out into rocks. The next year it accepted—it had little choice, really—the Moose fire from the Flathead National Forest, then 43,000 acres and let it burn itself out for another 28,000 acres until it hit the site of the old Anaconda burn and expired. The breakout year, however, came in 2003.

The fires grouped into seven complexes for a total of 145,000 acres, or 13 percent of the park; two of the complexes, the Robert and Wedge Canyon, burned 45,659 acres in adjacent national forests. The Robert rambled over the Apgar Mountains and Howe Ridge, the scene of the 1925, 1926, and 1929 Half Moon burns. The Trapper fire reburned much of the 1936 Heaven's Peak and the 1967 Glacier Wall fires. To protect against the Wedge Canyon fire, Parks Canada bulldozed a line 100 feet wide and 10 miles long just north of the international border. Fighting the fires would have been hopeless—there were too many fires in the region and too few suppression resources to throw at those in Glacier. The park felt instead that it could manage those it had, use the techniques it had pioneered on the Howling fire, keep the burns out of headquarters, and let them do their ecological work in the backcountry. Its neighbors were still evolving toward such notions. They all had different standards and National Environmental Policy Act approvals. In 2003 Glacier burned more acres than in any year since its creation.

Glacier had gotten its black, and then more. In 2006, despite attempts to suppress it from the onset, the Red Eagle fire blew up to 19,153 acres in the park before blasting out the park's eastern borders for another 15,050 acres on the Blackfeet Nation. By the time the 2010 fire management plan had coded the new strategies into bureaucratic language, some concerns were being voiced by resource management officers that there might be enough fire for a while, that the park could use a few years to assimilate what it had, that it might think about protecting some rare or old-growth patches in the name of biodiversity regardless of whether free-ranging fire, however natural, might be prepared to take them.

To some observers the value of a pause pointed to the future, that fire management needed to serve goals beyond getting the black, that this was not Big Burn National Park. To others, however, the real concern was not the fires recycling lodgepole pine or cleaning out western red cedar groves. It was that other big burn, the one that was driving out the glaciers.

———

Much of the Glacier National Park story is about boundaries. The park itself was created by slashing out a big chunk of the Blackfeet Reservation. Its northern perimeter traces the international border with Canada. It's flanked elsewhere by private holdings and national forests. These are not borders that matter to fire.

For its early years the park's fire problem was to keep fires that started outside from burning in. More recently it's to keep fires that start within the park from burning out. As national forests like the Flathead move away from extractive industry to recreation and wilderness, their aligned goals have allowed for more mutual fire plans. Park and forest can accept fires from one another and redraw their management strategies to work with fire behavior considerations rather than against them. The northern border delimits different thinking about how to restore fire. Under guidelines to enhance ecological integrity, Parks Canada favors prescribed fire, even high-intensity burns. Under notions of wilderness and "primitive America," the National Park Service leans toward letting natural fires, or loose-herded wildfires, do the job. The eastern border with the Blackfeet Nation is tougher since prevailing winds will drive fires—crown fires—from the park outward. Proposals surface from time to time to construct a fuel-break along the perimeter, but it would take a swath a mile wide to break the chain of wind-borne spots. Today, every entry into the park is burned. What a visitor first sees at Glacier National Park is a scorched land.[5]

The most interesting border is along the south, particularly the main ingress at West Glacier. The entry itself, through a small town on private land, is unburned. Behind it, however, at the visitor center and concession area, lies Apgar Mountain and the backdrop to Lake McDonald and Howe Ridge, and these were burned over in 2003. Yet the most significant fire scar lies outside the West Glacier complex: it's an overpass for the BNSF Railway, which visitors must drive under to enter the park. Its predecessor, the Great Northern, was instrumental in promoting the park, or in other words, the park as park is partly an artifact of steam. It's not simply academic muttering to suggest that this industrial gateway too is a portal of fire.

The eponymous glaciers for which the park is named also represent a blurry border, in this case of climate or of climate inscribing geologic epochs. The end of the Pleistocene, and the onset of the Holocene, is usually reckoned at between 10,000 and 11,500 years ago when the latest interglacial became undeniable. But the fundamentals behind the ice ages did not change. The Sun radiated as it had before; the Milankovitch cycles, with their stretching, tilting, and wobbling, continued; no replumbing of Earth's heat transfer machinery occurred equivalent to Pleistocene-onset shutting of the Isthmus of Panama. By geologic standards the Holocene is simply another warm period within a 2.6-million-year-old climatic wave train of chillings and warmings. The ice should return. In fact, by some accountings, it should have already begun to mound up. Instead of still-shrinking glaciers we should be witnessing spreading ones.

We don't, because the ice didn't. The Little Ice Age did not swell into a full-blown glaciation. Instead, the planet has warmed. It continues to warm in defiance of geologic history. The evidence grows that the cause for the warming is humanity, or more precisely, a change in how Earth's keystone fire species conducts its business. We shifted from a primary emphasis on burning surface biomass to burning fossil biomass. That pyric transition has taken vast amounts of stored lithic carbon and released it as gases. The Earth could adapt to changes in the rhythms and scale of burning grasses, shrubs, and forests; it has done so for over 400 million years. It could not adjust to the sudden upheaval in planetary combustion, or rather, it is undergoing an accommodation that will take centuries. So remarkable and recent has this revolution been that many observers believe the contemporary scene deserves its own geologic epoch, the Anthropocene, which begins in the late 18th century when fossil fuel combustion becomes more than locally significant. (For that reason it might as well be termed the Pyrocene.) For fire history this redrawing of the geologic timescale is akin to the arbitrary geographic bounding of places like Glacier National Park.

The park's mountain glaciers are shriveling. The USGS estimates that the original site held 150 glaciers in 1850, and that in 1910, when the park was formally founded, all were still present. In 2010 only 25 glaciers larger than 25 acres still existed. Some prognoses suggest that all the permanent ice will be gone in another 50 years. As a scenic spectacle and geologic presence, the collapsing glaciers are being replaced by conflagrations. The ice age is yielding to a fire age.[6]

The two epochs differ, and not just in their temperatures. The propagating fire is not simply an inverse of the melting glacier. Glaciers rise and fall according to purely physical conditions. For fires to shrink or spread the ambient temperature, or climate generally, must refract through ecological systems and human societies. Still, ice and fire can't coexist. As the ice recedes, plants will reclaim that land, and as conditions favor, they will burn. There are few ways to stanch the retreat of the glaciers. There are many ways to intervene in the prospect of burning woods. The future of fire management at Glacier will likely hinge on how to understand that pyric border and manage across it. Wildland fire management can do little with the deep drivers; it can no more halt fossil-fuel combustion than it can the sprawl of development behind the wildland-urban interface. It can, however, cope with the consequences.

———

The fire economy of the Northern Rockies is one of episodic big burns in which capabilities are almost inevitably out of sync with needs. There are too many pulaskis on hand during the slow years and not enough during the peak ones. Recurring crises push institutions to move responsibility up the chain of authority; as with wars or emergencies, big fires tend to centralize power. Yet the issue is no longer how to stop such fires but how to live with them and in places restore them in ways that help the habitat without burning down human settlements. The thrust is toward landscape-scale management, a geography larger than the biggest burns.

In the future, the park must think similarly across temporal scales. The real economy of fire, like the dual portal at West Glacier, must include internal combustion as well as free-burning flame. The combustion of fossil fuels is the big burn that will shape the future. How to draw boundaries around it is far from clear.

EPILOGUE

The Northern Rockies Between Two Fires

THE NORTHERN ROCKIES are an oddity: a small place politically but full of big fires with large political consequences. The region is distinctive enough to define a physiographic province, but too small, too sparsely inhabited, and too split among states to become a political power. Most of its lands are in federal hands, but even those are parsed among multiple agencies, national forests, and three Forest Service regions—four if you consider outliers like the Blue and Wallowa Mountains. Yet what fights against this fragmentation of influence is the region's singular concentration of wilderness, both legal and de facto. The wild has bestowed a cultural interest greater than empty land, and it allows the federal government to fill what would otherwise be an institutional void. The region punches above its demographic weight.

Excepting Yellowstone, always a special case, the regional politics, and the interest to national politics, follows the money, which aligns with state boundaries. The classic survey of Montana—Joseph Kinsey Howard's *Montana: High, Wide, and Handsome* (1943)—identifies the state with its vast plains, which were destined to replace its mountain minerals with coal and gas. Idaho aligns with its southern irrigated farmlands within the living periphery of Deseret, the old Mormon culture region. Wyoming is too dispersed for much beyond livestock and drilling rigs. The three states have a collective population less than that of Los Angeles—the city, not the metro region. In terms of population density they rank 44, 48, and 49 among the 50 states.

The other fire culture regions have populations to give them heft. California, for the all-hazard emergency service; Florida, for prescribed burning; Texas, for institutional minimalism—rank first, fourth, and second for population among the states. That grants their fire programs political punch. The Northern Rockies have nothing like that clout, and have invested what political muscle they have in establishing wilderness. Yet the Northern Rockies have influenced national fire policy from the time that big burns morphed from those associated with landclearing around rural settlements to those lodged in permanent wildlands. The conflagrations of the Northern Rockies have sent their smoke, and made their influence felt, far beyond the Cordillera.

The region's demographics stalled as extractive industries matured, then self-consumed themselves away. Mines, commodity agriculture, timber— each boom led to a bust or stagnation. Without a manufacturing base or a local construction market, state populations flattened, or even contracted. When the economy revived toward the end of the 20th century, it shifted to services and tourism, and its demographic surges became seasonal rather than permanent. What had been a regional embarrassment or liability, vast lands without permanent inhabitants, became an attraction for visitors, for environmental controversies, and for fire management.

Cultural interest, too, deepened. The region's self-identity had long fixated on Charlie Russell's premodern ideals or the first encounters of explorers and trappers on the model of Lewis and Clark, Osborne Russell, and Andrew Garcia; they spoke to a natural man in a state of untrammeled nature. As the modern economy began to focus on recreation and wilderness tourism, those renditions were retouched, updated, and recycled as a celebration of the wild. Fire programs followed suit. They began with fiery encounters that overwhelmed the newcomers; then underwent a hard phase of protection, or the ecological equivalent of an extractive economy; and finally eased, full of sound and fury, into an accommodation in which management meant a light hand on the land, and ultimately the illusion of lands untouched. The landscapes become valued to the extent that they are not formally settled.

Traditionally, the greater the human investment, the greater the human interest. In *Landscape and Meaning*, his meditation on how landscapes become culturally valued, Simon Schama imagined meaning free-associating and quantum-leaping from one signifier to another until the most valued lands of Earth seem like a self-referential art gallery or echo chamber of human artifice. The paradox of the Northern Rockies is that the less obvious the human investment, the deeper the social interest. In the status orderings possible among nature, the wildest achieves here the highest ranking.

This matters because fire practices ultimately derive from culture. However much science might be rallied to bolster ideals and social wishes, the values are stirred into them so completely, like sugar dissolved into coffee, they cannot be separated. They will drive the science, not the science the norms. In the Northern Rockies what began as a traditional frontier to be developed into garden and factory sank instead into a void bypassed by settlement until it seemed to booster a vague embarrassment to be hidden from view. Then, reversing the trend outlined by Schama, in which associations pile one on another, thickening the cultural deposition like sediments of meaning, the culture picked up the other end of the yardstick and began to valorize the vacant over the settled precisely because it was not, visibly, an expression of artifice. Later generations forgot that history of deliberately vacating and focused on what seemed to them as untrammeled lands that appeared to transcend history and culture.

More sharply than in other places fire history in the Northern Rockies is a dialectic between choosing to act and choosing not to act. The acting part—fighting fires, mounting complex campaigns to battle blazes in the backcountry—has a proud, easily understood narrative. There is the enemy, there is the hero, here the saga of their conflict. The not acting part—choosing to hold the land for the future, choosing to bar logging, mining, ranching, subdividing—lacks an equivalent power. The plot pivots on a strenuous inactivity. The moral core involves a decision not to engage. It means granting fires room to roam. It means withholding the airtankers, the helitackers, the smokejumpers, the hotshots, the engines. A hero is judged by the strength of his antagonist: the blowup fire makes a marvelous adversary. But a hero who stays his hand, who refuses the apparent challenge, who regards the thrown gauntlet as an invitation to study rather than leap to his feet, doesn't fit the templates. It's the fire community's problem of the Christian hero—how to take the epic, a literary form written to celebrate pagan values, particularly a warrior ethos, and rewrite it to celebrate turning the other cheek, to let the meek rather than the mighty inherit the Earth. The focus is the firefight not staged; the resolution is something that doesn't happen; the narrative more resembles a bank run that doesn't take place than one that convulses in panic. Catastrophe and ruin, particularly if openly confronted, offer the more easily engaging storyline.

For much of the 20th century the region and the national fire community knew how to tell the story of big burns. But then that backcountry became wilderness, and fire was not something that threatened the land with ruination but a chance to revive and restore it, and the inherited narrative no longer worked. What replaced it was not an epic of wilderness fire but a tragedy of young men and fire. The firefight remained the core action. But while Norman Maclean transformed a story of disaster into one of genuine tragedy, the real crisis at Mann Gulch was that there were smokejumpers there at all. They were as destructive to the land as big-tree loggers, fur trappers, and copper miners. That fire belonged as much as grizzlies and bighorn sheep. The Northern Rockies lost one opportunity to engage the new era when it failed to have the conversation it needed during the Yellowstone fires of 1988. It failed again when it declined to discuss the other tragedy at Mann Gulch. No new story of comparable power arose to replace the Big Blowup, only a more sophisticated retelling of its basic theme. Now, as wilderness joins other landscapes caught up in the maelstrom of the Anthropocene and its megafires, it will have a third chance to transmute the big into the significant.

In the Northern Rockies there are two parallel chronicles of big burns. One traces flames within wildlands. The other, invisible in its various combustions, burns fossil biomass in machines. The first recycles forests, prairies, and landscapes. The second restructures societies in the way they use and value the wood, grass, and scenery of those settings. Since the recession of the ice, the first has shaped how fire appears in the mountains. Since the onset of industrial combustion, the second has increasingly shaped how humanity tends, replaces, amplifies, or suppresses those landscape flames. Unlike geometric parallels, these often cross, though no narrative exists to map their braiding channels. Look to Glacier National Park for a cameo.

The region has plenty of big burns in the mountains and ample fossil fuels buried beneath the plains. One point of crossing is that industrial fire makes possible the society that values the wild as wild. This is not simply a matter of wealth, but of how the shift in combustion economies underwrites the way a society lives on its land. Coal, oil, and gas replace other extractive industries. It's not an accident that pristine nature protection (and in America, the idea of wilderness) has advanced stride by stride with a fossil-fuel civilization. The requirement that wilderness be roadless allows the two landscapes to segregate into distinctive realms.

But there are other links that join them as well. The use of internal combustion machinery to control free-burning fire in wildlands. The overloading of the atmosphere with greenhouse gases that seems to be leveraging big burns into mega burns. A tourist-and-recreational economy that relies on gas-powered vehicles to transport visitors from afar. When the time came to celebrate the 40th anniversary of the Bad Luck fire, the first natural fire in the national forest system, participants inspected the Selway-Bitterroot Wilderness from an airplane. In this regard the Northern Rockies stand to the country as wilderness patches do in the region.

What does not exist is a narrative to bond the two realms of fire under a common theme. The prevailing stories are still firefights: they are what distill character, conflict, and plot. Yet the great contribution of the region to the national fire saga and to a global fire philosophy of theory and practice is precisely the fire left as much as possible to free-burn. The Northern Rockies is that conception's heartland; not just where it was tested, but where it has fitfully thrived, spared extinction along with brown bears, and from where it can spread in the form of managed wildfire. There are long-lived programs elsewhere but none with the scope offered here. The others are isolated on sky islands or high-elevation granite massifs, or break out here and there as boutique burns, or burn in Alaska, where for practical purposes they might as well be in a separate country. Outside the Northern Rockies, apart from Alaska, they are, however boldly announced, a niche practice. Fire management will rely instead on prescribed fires or loose-herded wildfires to promote the black they want. They will avoid, where possible, landscape fires that may be many times larger than the legal wilderness.

The notion and its expression on the ground will survive, or expire, in its Northern Rockies hearth. This also means that wilderness fire, as a natural fire, will not spread much beyond those few sites where the wild is expansive. But it may not have to. It may be enough to have ample demonstrations of what a quasi-natural regime might look like; to show, in the field, how such fires behave and so make it possible to calibrate models and fine-tune field operations. The real payoff may be in underwriting the contemporary doctrine that "fire is fire" and using fires of opportunity—whatever fires occur from whatever source—to get the burning done. Such techniques would not be possible (or acceptable) without the exemplar of the wilderness fire. They distribute by secondhand not only the experience but the grace of the wild.

Now the Northern Rockies may be poised to segue into another era of big fires on big lands. The first generation of wildland fire had conservation as it goal. It sought to eliminate waste and promote multiple use and considered that fire control was a precondition of both. Given the havoc at the end of the 19th century that ambition had its logic. The firefight, from the Great Fires to Mann Gulch, provided the core narrative. The second generation shifted from custodial conservation into outright preservation, best epitomized by the Wilderness Act. It sought to restore good fire, and believed, for a while, that sheer size might compensate for lapses in knowledge and technique. No narrative of comparable power emerged; on the contrary, the masterwork returned to the firefight in a remote landscape. The emerging third generation must wrestle with sustainability and resilience in the face of global environmental upheavals, however that comes to be understood within the inherited public lands. It won't mean big-screen firefights, because there is neither will nor money to do them, outside of community protection. It won't mean restoration by prescription because the required scale is too vast and deliberate burning may well be deemed a violation of the legal character of the Wild. The coming age will replace multiple use with mash-up and big blowups with megafires, but whether it finds a way to tweak the old narrative with new purpose is unclear. It needs a voice to match its vision, and a structure to hold its story.

===

English Creek, Ivan Doig's novel of Montana in the 1930s, after the West was won and before it became a tourist destination, is a hardscrabble family saga, but one whose inflection points always seem to align with fires. One reason is that the central character, Varick "Mac" McCaskill, is a district ranger, which in the Northern Rockies means having to deal with lots of fires and with the more than occasional big one. But there could be few plot devices drawn from nature with as much dash and tension as a firefight.

The fires become a rite of passage for each generation. They make men out of boys; break men whose character isn't up to the ordeal; make and break friendships; define the dominant institution on the landscape, the U.S. Forest Service, which in the middle of the decade announces the 10 a.m. policy as a means to halt the serial conflagrations. The chronicle effectively begins with the Big Blowup. "That same 1910 smoke never really left my father," explains Jick McCaskill, the 14-year-old who tells the tale. The memories stayed "in my father, smears of dread." That was the yardstick, though each generation had its own updated reburns. "I had grown up hearing of forest fires. The storied fire summers, Bitterroot, Phantom Woman, Selway, this one, they amounted to a Forest Service catechism." Then Jick got to see one up close and personal, and it is decisive. The climax comes with a daring firefight and an even more desperate burnout. Later, an adult returned from World War II, Jick joins the smokejumpers and then the Forest Service before turning to ranching.

English Creek is not by theme a fire book. But it would be hard for a realist novel about the region, particularly in those years, not to have fire in its pages—this one has them calibrating the chronicle the way a town tower's bells record the passing hours. What is surprising is that so few writers have heard them chime. Then, eight years after *English Creek*, Norman Maclean recorded their sound in unforgettable cadences. But he, too, looked back on a heroic or at least a silver age of bold firefights and Shakespearean-scale tragedy.

There is a pivotal scene in the Mann Gulch tragedy when, having witnessed a blowup in the making and coming at them, Wag Dodge orders the crew to drop their packs and heavy tools. Some had already discarded them, some held them to the end, and commentators have taken that moment as critical to their sense of what was happening since without their tools they didn't know who they were or what they ought to do. From the origins of wildland fire protection, its tools have told its community who they are; the enduring symbol of the Big Blowup, after all, is the pulaski. Dodge himself evades the issue by striking a gofer match, lighting an escape fire, and seemingly inventing a new tool for the occasion. When, in the early 1970s, fire crews were told to drop their tools, they hesitated, and not without cause. The prescribed fire as an escape fire—in this case, an escape from past history—was a novel and untested notion. Those big burns were still coming at them.

Over the past 40 years, although the fire culture of the Northern Rockies has learned to trade one tool for another, it has been reluctant to drop the old ones, and wary about following Dodge into a fire of his own making. Knowing what to drop and what to pick up, and when to do it, and how to do it with blowups at your back is no simple task. It is one that generations of young men and old have faced here, where the mountains roar, and one they will certainly have to keep relearning until that vast hole in the sky, as Norman Maclean once called it, seals shut.

NOTE ON SOURCES

B ECAUSE OF ITS ROLE in the national fire narrative, I knew many of the basics behind fire in the Northern Rockies and something of the nooks and crannies of its history and geography. For general background I relied on Michael P. Malone, Richard B. Roeder, and William L. Lang, *Montana: A History of Two Centuries*, rev. ed. (Seattle: University of Washington Press, 1991) and Leonard J. Arrington, *History of Idaho* (Moscow: University of Idaho Press, 1994), both of which helpfully included chapters on arts and letters. The mother lode, however, is government documents, and especially the veins running through the Missoula fire lab. Few of the essays in this collection do not have an adit into them at some point. The particulars are given in the notes for each essay.

Early on in the project that became *To the Last Smoke* I began alluding to literature as a way to avoid recycling data and repeating endless scientific arguments. The habit grew. I've found it enjoyable, and so found my way to William Kittredge and Annick Smith, eds., *The Last Best Place: A Montana Anthology* (Helena: Montana Historical Society Press, 1988), that astonishing watering hole of Montana oral lore and written texts. Mostly, though, I collected nuggets on the ground as I tried to follow the stream of my travels to their sources.

NOTES

WHY BOISE IS NOT THE NATIONAL CENTER FOR FIRE

1. Background history and quote from "Progress Report: Bureau of Land Management Fire Protection Program Planning and Development" (unpublished report, BLM, 1969), 3–4. Jack F. Wilson, "From Whence They Came: A Perspective on Federal Wildland Firefighters in the Department of the Interior" (unpublished manuscript, copy given to author by Wilson), 6–7. On emergency presuppression funds, see Samuel M. Brock, "BLM Fire Control Study: Economics of Fire Control with Special Reference to BLM Protection Operations" (unpublished report, 1964), 21. For a fuller survey of the context, see Stephen J. Pyne, *Between Two Fires* (Tucson: University of Arizona Press, 2015)

THE PARADOXES OF WILDERNESS FIRE

1. Hugh Trevor-Roper, "The Invention of Tradition: The Highland Tradition of Scotland," in *The Invention of Tradition*, ed. Eric Hobsbawm and Terence Ranger, Canto edition (Cambridge: Cambridge University Press, 1992), 15, 24.
2. References and quotes from Susan L. Flader and J. Baird Callicott, eds., *The River of the Mother of God and Other Essays by Aldo Leopold* (Madison: University of Wisconsin Press, 1991), "The Maintenance of Forests"; "To the Forest Officers of the Carson"; "Conservation in the Southwest"; "Grass, Brush, Timber, and Fire in Southern Arizona"; "Wilderness as a Form of Land Use"; "Wilderness"; "Conservationist in Mexico." Not in the collection is "'Piute Forestry' vs. Forest Fire

Prevention" (1920). For that, see David E. Brown and Neil B. Carmony, eds., *Aldo Leopold's Southwest* (Albuquerque: University of New Mexico Press, 1990), 139–42.

FIRE'S CALL OF THE WILD

1. Thanks to Bob Mutch and Dave Campbell for inviting me to join their anniversary flight, and then sharing over the course of an afternoon their deep experience with the SBW. It goes without saying that the opinions expressed in this essay are mine alone.

2. On 1967 fires, William R. Moore, Assistant Director, WO-Fire Control, "Report on Field Travel in Region 1, August 8–19 and September 3–9, 1967," National Archives, Record Group 95, Accession 72-A-3046, Box 168, p. 6. On "practically illegal," from Linda S. Mutch and Robert W. Mutch, "Wilderness Burning: The White Cap Story," unpublished manuscript with transcriptions from recordings from the anniversary event, courtesy of Robert Mutch.

FIRE BY PARALLAX

1. Nicolas Point, *Wilderness Kingdom: Indian Life in the Rocky Mountains, 1840–1847: The Journals & Paintings of Nicolas Point*, trans. Joseph P. Donnelly (New York: Holt, Rinehart, and Winston, 1967).

2. On tribal history I follow John Fahey, *The Flathead Indians* (Norman: University of Oklahoma Press, 1974).

3. Fahey, *Flathead Indians*, 95.

4. On the establishment of the Forestry Branch, see J. P. Kinney, *A Continent Lost, A Civilization Won* (Baltimore: Johns Hopkins Press, 1937), 249–80.

5. On the 1910 fires, see Historical Research Associates, *Timber, Tribes, and Trust: A History of BIA Forest Management on the Flathead Indian Reservation (1855–1975)* (Dixon, MT: Confederated Salish and Kootenai Tribes, 1977), 43–45, 238–39.

6. Seasonal cycle paraphrased from Fahey, *Flathead Indians*, 78–79.

7. Fire episodes from Fahey, *Flathead Indians*, 172, 201; 1910 reference from HRA, *Timber, Tribes, and Trust*, 238.

8. I would like to thank Jim Steele and Tony Harwood for generously sharing a morning of their busy days to explain how a change in spectacles can change the spectacle being seen, and then for sharing maps and

documents. I could not have written this piece without them. I should not need to add that the resulting text can indeed be read several ways.

9. For the basics of the legislation, I rely on Donald L. Fixico, *Bureau of Indian Affairs* (Santa Barbara, CA: Greenwood, 2012).

10. An excellent website has cached the relevant documents, including the 2007 fire management plan, which, unusually, includes explanations of the thinking behind decisions, not just the protocols to apply those decisions. The website keeps changing, however, so visit http://www .cskt.org, then link to various fire topics, as desired.

THE EMBERS WILL FIND A WAY

1. This profile could not have been written without the help of Jack Cohen, despite his unintended efforts in person and in print to overwhelm me with a swarm of data. He left the wording, the conceptual insights, and the organizing conceit to me, which is to say any errors are mine. It was a pleasure to listen to him and to feel his radiant enthusiasm, at least some of which I hope has found a glow in my sketch.

YOUNG MEN, OLD MEN, AND FIRE

1. Bertrand Russell, *The Scientific Outlook* (New York: Norton, 1931), 101–2; Bud Moore, *The Lochsa Story: Land Ethics in the Bitterroot Mountains* (Missoula, MT: Mountain Press, 1996), 3–12. Other information on Moore also from two oral histories, one by USFS for "The Greatest Good" and one by Jamie Lewis for the Forest History Society, both available through the Forest History Society.

2. William R. Moore, "Towards the Future . . . Land, People, and Fire," *Fire Management* 35, no. 3 (Summer 1974): 5.

3. I need to thank Bob Mutch for sharing his career during a long conversation and for graciously allowing me to poach on his own literary plans to describe his experiences in the SBW.

HOW I CAME TO MANN GULCH

1. I wish to thank Craig Kockler of Helena National Forest for sparing a few minutes at the end of a long day to pass along information regarding the Meriwether fire.

WHAT MAKES A FIRE SIGNIFICANT?

1. Stephen W. Barrett, Stephen F. Arno, and James P. Menakis, *Fire Episodes in the Inland Northwest (1540–1940) Based on Fire History Data*, General Technical Report INT-GTR-370, U.S. Forest Service, 1997), 3.
2. Incident reported in the *Brooklyn Daily Eagle*, August 14, 1889.
3. Reported by the *Sacramento Daily Record-Union*, Sacramento, CA, August 2, 1889.
4. *New York Times*, August 24, 1889.
5. Telegraph reference from *Lawrence Daily Journal*, July 30, 1889.
6. Wallace Stegner, "A Sense of Place," in *Where the Bluebird Sings to the Lemonade Springs* (New York: Modern Library Classics, 2002), 205.

THE SECOND BIG BLOWUP

1. This essay forms the foundation for a subchapter in *Between Two Fires*. I include it here at the risk of repetition because the 1988 fires were a significant event in the Northern Rockies and I could neither ignore them nor invent another essay to replace this one.
2. The literature on the fires is huge. I relied on the following for the basics: Ronald E. Masters, et al., eds., *The '88 Fires: Yellowstone and Beyond Conference Proceedings* (Tallahassee, FL: Tall Timbers Research Station, 2009); Rocky Barker, *Scorched Earth: How the Fires of Yellowstone Changed America* (Washington, DC: Island Press, 2005); Linda A. Wallace, ed., *After the Fires: The Ecology of Change in Yellowstone National Park* (New Haven, CT: Yale University Press, 2004); and Hal K. Rothman, *Blazing Heritage: A History of Wildland Fire in the National Parks* (Oxford: Oxford University Press, 2007). For a good distillation, see Kathleen M. Davis and Robert W. Mutch, "The Fires of the Greater Yellowstone Area: The Saga of a Long Hot Summer," *Western Wildlands*, Summer 1989, 2–9. Beyond that, I deferred to my summer of fire planning in 1985.
3. Estimates from Rothermel and Despain from Barker, *Scorched Earth*, 205.
4. To clarify my position in the years leading to the 1988 bust, I decided to scan the dot-matrix printouts of my reports but I have not found a long-term website in which to park them. If a reader would like a copy, please write me.

THE OTHER BIG BURN

1. I want to thank Dennis Divoky, fire ecologist, for devoting an afternoon at the end of a busy week to a tutorial on Glacier fire, and to Deirdre Shaw, archivist, for help in identifying and copying a stack of relevant historical documents. Glacier is remarkable in having records, exemplary in its ability to access them, and exceptional in the willingness of its staff to make them available. They made me feel like a historian up to his elbows in paper rather a journalist dependent only on oral interviews.

2. Superintendent memorandum, November 30, 1967, "Public Relations and Press Coverage During the Fire Emergency" and "Proceedings—Glacier Forest Fire Review: November 30–December 1, 1967," both in Glacier National Park Archives; Box 309 Wildland Fire Management (Y14); Folder 21.

3. For quote, see p. 35.

4. Thomas Zimmerman, Laurie Kurth, and Mitchell Burgard, "The Howling Prescribed Natural Fire—Long-Term Effects on the Modernization of Planning and Implementation of Wildland Fire Management," in *Proceedings of 3rd Fire Behavior and Fuels Conference, October 25–29, 2010* (Spokane, WA: International Association of Wildland Fire, 2011).

5. See David L. Bunnell and G. Thomas Zimmerman, "Fire Management in the North Fork of the Flathead River, Montana: An Example of a Fully Integrated Interagency Fire Management Program," in *Fire in Ecosystem Management: Shifting the Paradigm from Suppression to Prescriptions, Tall Timbers Fire Ecology Conference Proceedings, No. 20,* ed. T. L. Pruden and L. A. Brennan (Tallahassee, FL: Tall Timbers Research Station, 1998), 274–79.

6. See USGS, "Retreat of Glaciers in Glacier National Park," last modified May 2013, http://nrmsc.usgs.gov/research/glacier_retreat.htm.

INDEX

ABOUT THE AUTHOR

Stephen J. Pyne is a historian in the School of Life Sciences at Arizona State University. He is the author of over 20 books, mostly on wildland fire and its history but also dealing with the history of places and exploration, including *The Ice*, *How the Canyon Became Grand*, and *Voyager*. His current effort is directed at a multivolume survey of the American fire scene—*Between Two Fires: A Fire History of Contemporary America*; and *To the Last Smoke*, a suite of regional reconnaissances, all published by the University of Arizona Press.